Skeptoid 3
Pirates, Pyramids, and Papyrus

By Brian Dunning

Foreword by Richard Saunders
Illustrations by Nathan Bebb

Skeptoid 3: Pirates, Pyramids, and Papyrus
Copyright 2011 by Brian Dunning
All Rights Reserved.

Skeptoid Podcast ©2006-2011 by Brian Dunning
http://skeptoid.com

Published by Skeptoid Media, Inc.
Laguna Niguel, CA

First Edition
ISBN: 978-1453881187
Printed in the United States of America

It ain't what you don't know that gets you into trouble. It's what you know for sure that just ain't so.

Mark Twain

Acknowledgements

Although I do nearly all of the work researching and writing Skeptoid on my own, the following individuals provided a massive amount of help with referencing and finding further reading suggestions for all the chapters in this book, and I am greatly indebted to them:

Michael Arter, Mike Bast, Erwin Blonk, Mike Bohler, Katie Brockie, Justin Crain, Adam Deso, Josh DeWald, Lee Dunn, Jack Flynn, John Folsom, Kevin Funnell, Derek Graham, Greg Hall, Kerry Hassan, Diane Johnson, James Lippard, Kelly Manning, Tom Markson, Bob McArthur, Mark Metz, Dennis Mitton, Lee Oeth, Leonardo Oliveira, Kathy Orlinsky, Tom Rhoads, Rico Sanchez, Tom Schinckel, Thomas Shulich, Mike Weaver, Glen Wheeler, and Sarah Youkhana.

Also thanks to Joe Nickell for giving me a (somewhat more than) gentle nudge improving the quality of these books.

To Erika –

The brightest light in my universe, and quite possibly in all the other ones as well.

Contents

Foreword: Why Skeptoid Is Part of My Life 1

Introduction 3

1. Search for the Missing Cosmonauts 5
2. How Dangerous Is Cell Phone Radiation? 12
3. It's Raining Frogs and Fish 18
4. The Battle of Los Angeles 24
5. Demystifying the Bell Witch 29
6. Falling Into Mel's Hole 35
7. The Truth about Aspartame 40
8. Scalar Weapons: Tesla's Doomsday Machine? 47
9. HAARP Myths 52
10. Ten Most Wanted: Celebrities Who Promote Harmful Pseudoscience 58
11. Betty and Barney Hill: The Original UFO Abduction 65
12. The Incorruptibles 71
13. The Oak Island Money Pit 77
14. Space Properties for Sale 82
15. The Bohemian Club Conspiracy 87
16. The Sargasso Sea and the Pacific Garbage Patch 94
17. Chasing the Min Min Light 99
18. The Rendlesham Forest UFO 105
19. Who Is Closed Minded, the Skeptic or the Believer? 114
20. The Angel of Mons 120
21. Is He Real, or Is He Fictional? 126

22. The Case of the Strange Skulls.................................... 132
23. Kangen Water: Change Your Water, Change Your Life 140
24. The Bosnian Pyramids 146
25. The Lucifer Project 153
26. FEMA Prison Camps 159
27. How Old Is the Mount St. Helens Lava Dome?........... 165
28. Binaural Beats: Digital Drugs..................................... 171
29. Coral Castle.. 177
30. The Placebo Effect 183
31. Attack of the Globsters!... 189
32. Was Chuck Yeager the First to Break the Sound Barrier?
.... 195
33. NLP: Neuro-linguistic Programming 203
34. The World According to Conservapedia.................... 209
35. High Fructose Corn Syrup: Toxic or Tame?............... 215
36. The Mothman Cometh.. 221
37. Sarah Palin is Not Stupid 226
38. Locally Grown Produce.. 232
39. How to Make Skepticism Commercial 238
40. What's Up with the Rosicrucians?.............................. 244
41. Real or Fictional: Food and Fashion 250
42. Organic vs. Conventional Agriculture 257
43. Should Science Debate Pseudoscience?..................... 265
44. Decrypting the Mormon Book of Abraham............... 271
45. Should Tibet Be Free?... 277
46. How to Be a Skeptic and Still Have Friends.............. 283

47. Daylight Saving Time Myths	288
48. All About Astrology	294
49. More Medical Myths	303
50. Shadow People	309

Foreword: Why Skeptoid Is Part of My Life

by Richard Saunders

Fourteen hours... not a long time really, unless you are flying from Sydney to Los Angeles on a clapped out 747, then it's a very long time indeed. As the night hours dragged and the inflight movie ended on yet another trip to the United States, when all hope of getting sleep had vanished, I put on my eye mask and turned to my one true friend...my iPod, and heard those familiar words. "You're listening to Skeptoid. I'm Brian Dunning, from Skeptoid dot com."

We often hear that skeptics spend too much time preaching to the converted or preaching to the choir. I never find this a waste of time as I am a member of that choir and always appreciate more information and inspiration. In fact I need it. For me, that normally means climbing aboard those 747s and flying off to some of the major international skeptical conventions such as The Amazing Meeting (TAM) in Las Vegas held by the James Randi Educational Foundation or, more recently, Dragon*Con in Atlanta. There is nothing like hearing firsthand from the likes of James Randi, Dr. Phil Plait, Dr. Steve Novella and of course, Brian Dunning. But now, at your leisure, you can pour over some of the many topics Dunning has not only thoroughly researched but also formulated in to bite size and very readable explorations into the sometimes strange world of beliefs and investigations.

Often I find myself reaching for copies of Dunning's books and his podcast of the same name when researching skeptical topics for articles I'm called upon to write or to prepare for radio and TV interviews. I imagine that I am not alone in this. All over the world people are at once discovering and building upon their skeptical knowledge with the help of Skeptoid in all its forms. Without exception I learn a little more or even a great deal more from each new contribution. More than once the information I have gleaned from Skeptoid has stopped me from making an ass of myself when talking about the paranormal and such like. There is nothing worse for our way of thinking than the blathering of an uninformed or ill-informed skeptic.

Now we come to your role in Skeptoid, and a vital role it is to be sure. If you find that Dunning has lapsed in his research or made an error in his reasoning, let him know. At heart Skeptoid is following in the tradition of science and that tradition demands that research be questioned and conclusions be tested when and where necessary. It is no insult to science to be corrected from time to time, in fact science itself would not work without this measure. It is a testament to Dunning's sincerity that he will correct and update his findings in line with that tradition. Without doubt he is what we call in Australia, 'a fair dinkum bloke.'

In 1995 Bill Gates wrote *The Internet Tidal Wave* memorandum to Microsoft executives in which he outlined the impact of this new form of communication and directed that it be given the "...highest level of importance". From what we in the skeptical world have seen, Dunning has jumped on his surfboard of reason, paddled out into the Internet ocean and is riding the tidal wave for all it's worth. I hope he never wipes out.

I really don't know when Dunning finds time to sleep, let alone do all the research needed to write and record Skeptoid. But thank your lucky stars, your lucky shoes and your lucky socks he does.

Richard Saunders is one of Australia's leading skeptical investigators and communicators. A former President of Australian Skeptics, he is also the producer of the popular Skeptic Zone podcast. An author with around 30 books to his credit, Richard delights in travelling the world to teach his other passion, Origami.

Introduction

Chances are you're wondering what the heck kind of a book this is. The chapters seem to be all over the map; a boggling array of unrelated topics, urban legends mixed with popular products and claims, even some philosophy and history with a smattering of hard science. What's the glue that binds it all together?

These chapters are all adapted from episodes of my audio podcast, *Skeptoid*, available at Skeptoid.com. It's a show about examining the stories we hear in our daily lives, and how to analyze them to see what's fact and what's fiction.

There is no shortage of fiction being foisted upon us by the mass media. What masquerades as news and even as documentaries are often mere sensationalism, driven by the business model of needing to attract a large audience. It's easy to excite people with sensationalism. Tell people that something inexplicable and magical is going on, and you'll turn heads. The result is twofold: First, the TV networks, magazine publishers, and email spammers are happy; and second, the general public's collective knowledge level drops to the point where nobody cares what's real and we have this skewed sense of how our world actually works.

Decisions are based on this skewed information. Policy is formed. Voters make their selections. Entertainment companies pick their lineups. Healthcare and life-or-death decisions are made. People make important life decisions based on the quality of the information they have.

A growing number of people – loosely called skeptics, thus the title of this book and the podcast – are concerned about this. We demand a minimum standard for quality of information. We believe it's important to understand the way the universe actually works, to best prosper within it. The chapters in this book look at fifty topics, some of which you've heard of, others you haven't, but all are examined from a skeptical perspective. You'll see the thinking process behind the analysis of popular claims.

This process often puts us at odds with snake oil salesmen, with promoters of the supernatural and paranormal, and with providers of alternatives to healthcare that have failed to meet scientific standards. And, nine times out of ten, it puts us in the same camp as "the establishment" of scientific consensus and government regulators. To many people, this raises suspicion. I've been accused of being "on the payroll" of Big Government, Big Oil, Big Pharma, Big Science, Big Toxins, Big Education, and just about everything else more times than I can count.

I hate the appearance of "toeing the company line" just as much as you do, but a closer look at the real demographics shows that, by no means, are the conclusions of skeptics mainstream. By any survey you want to name, the majority of people are mistrustful of science. Most people believe in ghosts and other spiritual beings. Most people purchase unproven healthcare supplements. Most people accept that strange things happen all the time that defy scientific explanation, without ever bothering to asks the scientists themselves.

But who has the time or knowledge to properly investigate every such question? Almost nobody. Does this mean that we're all left to fend for ourselves, guessing and applying our television knowledge to know what's real and what's not? Unfortunately, it does, in most cases. The purpose of this book is to offer you some of the tools that will help you be right far more often than you're wrong. It's the value of critical analysis, and I think you'll find it a fascinating experiment.

– Brian Dunning

1. Search for the Missing Cosmonauts

During the late 1950's and early 1960's, the space race between the United States and the Soviet Union was hot. Both sides built and tested rockets as quickly as they could, trying to be the first to launch an artificial satellite into orbit, often with explosive results. Both sides had their successes, and both sides had their failures. People around the world watched and listened. Some, most notably amateur radio operators, listened more closely than others. And of these, a pair of young brothers from Italy, Achille and Giovanni Judica-Cordiglia, reigned supreme. Their library of audio recordings of nearly every flight from the space race is by far the most comprehensive private collection  known. But the real reason it's notable is that includes a number of recordings of alleged events that didn't make it into the history books: doomed Soviet cosmonauts captured in their final moments of life, on flights that the Soviets said never happened.

During the cold war, the Soviet Union was a knot of state secrets. More than anything else, the cold war was a war of propaganda, each side trying to show the world that they were the smartest, the fastest, the highest, and the best. In this context, it's not surprising at all that the true progress of their space program would be closely guarded and only the best news released to the world. With their state-controlled media, the Soviets had the ability to accomplish coverups of failures to a degree that would never have been possible in the United States.

Achille and Giovanni were creative and scientific geniuses in the truest sense, both in their twenties. When the Soviets announced the successful launch of Sputnik I on October 4, 1957 and published the radio frequency for everyone to hear, the brothers scavenged what radio equipment they could and tuned it in.

From that one recording, their self-taught education proceeded like a rocket (pun intended). They learned how to detect the Doppler effect in signals from orbit, and how to calculate an object's speed and altitude from that. They filled logbooks with conversion tables and Soviet frequencies. And so, when the Soviets launched Sputnik 2 only a month after Sputnik 1, they were well prepared. And this time, the brothers discovered something new: a pulse.

It was the heartbeat of Laika, a small dog. Sadly for Laika, Sputnik 2 was a one-way trip; there was no provision for reentry or recovery. Three months later, the United States launched its first satellite, Explorer I, and like the Soviets, published the frequency of the signal. Achille and Giovanni captured it, and then their lives as local media celebrities began. They were the darlings of the local papers and radio stations. They took over a nearby concrete bunker left over from World War II, made improvements to their equipment, and built larger antennas. They called their little radio observatory Torre Bert, and anytime anything launched into space from anywhere, Torre Bert was filled with friends, reporters, local scientists, and anyone who wanted a good time.

The Torre Bert experiment took a more serious turn on November 28, 1960. A West German observatory announced that it was receiving a strange signal on a Soviet space frequency. The brothers tuned in, and heard hand-keyed Morse code repeating the international distress signal, S-O-S, over and over again. Their Doppler calculations showed almost no relative speed, which they interpreted to mean that the distressed spacecraft was on a course directly away from the Earth. The signal grew weaker and was never heard from again. Apparently, the brothers had just recorded evidence that a manned Soviet spacecraft somehow got off course and left Earth's orbit, permanently.

About two months later in February 1961, variously reported as the 2nd or the 4th of the month, they picked up another transmission from space, which experts interpreted at the time as the dying breaths of an unconscious man.

And another signal from the same flight, interpreted by the brothers' father, a cardiologist, as a failing human heartbeat.

The brothers' story and recordings were played throughout Italy. Two days after this publicity, the Soviets announced the failed re-entry of a large unmanned craft.

In April of 1961, a journalist at the International Press Agency in Moscow tipped off the brothers that something big was about to happen. They turned on their equipment, and the next day, listened in on Yuri Gargarin's voice during the first manned space flight.

But, the most dramatic of the brothers' recordings came about five weeks later in May of 1961, the date variously reported as the 17th, 19th, or 23rd. A woman's voice transmission, translated as "Isn't this dangerous? Talk to me! Our transmission begins now. I feel hot. I can see a flame. Am I going to crash? Yes. I feel hot, I will re-enter..." (Note, this is the popular translation given on the Internet. I had some Russians of my own translate it for me, and her words were much less dramatic. Mostly counting off numbers, and nothing about flames, crashing, or re-entry. But it was recorded on a known Soviet space frequency.)

When I first heard about the Judica-Cordiglia recordings from Torre Bert, I was definitely intrigued. It simply appears plausible. We know that the Soviets covered up their failures. We know that their launch record in those days was absolutely abysmal, far worse than the United States. If Yuri Gagarin made it into space, it almost seems like a foregone conclusion that at least a couple of other guys must have previously died in the attempt.

Part of the trouble you find when you research this is that the recordings from Torre Bert are only one small square in a quilt the size of Texas. There are many, many stories circulating about missing cosmonauts who died in spaceflights as early as 1957. According to some Western intelligence sources, as many as 11 fatal Soviet accidents occurred, both in flight and on the ground, all before 1967. We know that the Soviets painted certain cosmonauts out of photographs, in fact you can see some great before & after examples of this on the LostCosmonauts.com web site. We know that the death of at least one cosmonaut killed in a training accident, Valentin Bondarenko was concealed until 1986, and even then was only declassified after western journalists found out about it in 1980. There's also considerable controversy about the case of Vladimir Ilyushin, who claims to have launched five days before Gagarin, but a problem caused him to re-enter early and land inside China, where he was held captive for a whole year. Some of Ilyushin's supporters even assert that Gagarin's flight never took place; rather that he was hastily shuttled to a mocked-up landing site in Ilyushin's backup capsule so the Soviet propaganda machine could attach a healthy, smiling young face to Ilyushin's heroic flight. Ilyushin still lives in Moscow at last report, and still maintains his story.

Much of the criticism of the Judica-Cordiglia brothers comes from space historian and author James Oberg, who wrote a book based on his investigations into all of these stories of lost cosmonauts. His principal conclusions were that there was insufficient evidence available to substantiate any of these stories. But Oberg's research concluded in 1973, when the Iron Curtain was still strong. 35 years later, virtually everything has been long since declassified. It's now possible to read detailed histories of those early days, and the dates and types of all their launches, failures included, is thoroughly documented. I compared the timelines of what the Judica-Cordiglia brothers recorded to the timeline of the Soviet space program. I did find some problems.

The main inconsistency is that during the times of the Morse code and the astronaut's alleged breathing and heartbeat sounds, the Soviets were still launching dogs and mannequins. A few days after the Morse code recording, Sputnik 6 carrying two dogs was deliberately self-destructed upon a failed re-entry, and three weeks after that, two dogs were launched and safely recovered even though the third stage of their Vostok booster failed and the craft did not achieve orbit.

While it's true that the Soviets did have a proven capability to escape the Earth by the time of the fading Morse code (Luna 1 had passed the moon a year earlier), the Vostok 8K72 booster only had the ability to lift 500 kilograms to escape velocity, way too small for a manned capsule. Even for several years afterward, the Soviets had no rocket capable of lifting a manned capsule beyond Earth's gravity.

In the two months following the brothers' recording of the breathing and the alleged heartbeat, the Soviets made two successful low Earth orbit flights, each carrying a small dog and a mannequin. These are the type of test flights made when you're not yet ready to launch a man.

Following the Soviets' success at launching Gagarin in April 1961, the Judica-Cordiglia version of events suggests that their next feat was to launch a woman, thus the May 1961 recording. However, the Soviets' next launch wasn't until August, and it was another man, Gherman Titov, who flew for a full day in orbit. Valentina Tereshkova, credited as the first woman cosmonaut, didn't fly until more than two years after Gagarin, in June of 1963.

Of course these inconsistencies don't prove anything, they just show that if you accept the Judica-Cordiglia assertions as fact, they show an illogical backwards progression by the Soviets that's contradictory with the character of the space race. The Soviets never took backward steps.

A more compelling reason to be skeptical of the Judica-Cordiglia brothers' interpretation of their recordings is the lack of corroborating evidence from the numerous, far more sophisticated radio tracking stations that existed. These were the days of the Distant Early Warning Line and the birth of the North American Air Defense Command, and the Americans, British, Canadians, Germans and French all had equipment that far exceeded the humble capabilities of homebuilt Torre Bert, with things like tracking dishes that Torre Bert lacked; and moreover, the western propaganda machine would have loved nothing better than to publicize Soviet failures. The best explanation for why such announcements were never made is that no such failures were ever observed.

Indeed, the story of the Soviets announcing a failed unmanned flight after hearing that the brothers' recorded their dying cosmonaut doesn't match the history books. It's a great sound bite but I found no such report anywhere. Moreover, current records show a successful test of an R7 booster carrying a dummy missile warhead on February 7, 1961, about the day of the claimed admission.

Am I saying Achille and Giovanni were hoaxers? Far from it. In fact, in researching their story, I gained tremendous respect for their abilities and for what they accomplished. As I said before, their library of recordings is a treasure of inestimable value, and there's a documentary film about them called Space Hackers, which I found on YouTube, and which I highly recommend. Unfortunately their story is too often told without critique or inquiry into the plausibility of their most extraordinary claims. There are simply too many other possible explanations for their recordings to comprise useful evidence of lost cosmonauts. Is there stuff we still don't know about the Soviet space program? Absolutely. Might it include accidents, even deaths? Probably. Might it include unknown spaceflight failures, possibly even lost cosmonauts? Maybe, but now you're into territory that western intelligence too easily could have known about. I maintain an open mind on the question.

References & Further Reading

Abrate, G., Abrate, M. "Erased from Memory." *The Lost Cosmonauts.* The Lost Cosmonauts, 24 Apr. 2004. Web. 1 Aug. 2008. <http://www.lostcosmonauts.com/erased.htm>

Burgess, Colin, Hall, Rex. The first Soviet cosmonaut team: their lives, legacy, and historical impact. Chichester, U.K.: Praxis, 2009. 203-228.

Haimoff, Elliot H. "Letter by Dr. E.H. Haimoff of "Global Science Productions"." Letter by Dr. E.H. Haimoff of "Global Science Productions" to the "My Hero" website where an account of Ilyushin's space mission by Mr Paul Tsarinsky was recently posted. Global Science Productions, 7 Aug. 1999. Web. 1 Aug. 2008. <http://abrax.isiline.it/servizio/letter.htm>

Oberg, James. *Uncovering Soviet Disasters.* New York: Random House, Inc., 1988. 156-176.

Zheleznyakov, Alexander. "Gagarin was Still THE First. Part Two." *Spaceflight.* 1 Nov. 2002, Volume 44, Number 11: 471-475.

2. How Dangerous Is Cell Phone Radiation?

Today we're going to pick up virtually any consumer magazine or open any Internet news web site and read about a frightening new threat: That radiation from cell phones is dangerous, perhaps causing brain tumors or other cancers, maybe even cooking your brain like an egg or like popcorn. Most people have no knowledge of science other than what they hear on the news, so we have a whole population growing up with this understanding. Is the fear justified? Do cell phones have the potential to cause physical harm, or are they completely safe? Or, like so many other questions, is the truth somewhere in the middle?

Let's take a closer look at exactly what kind of threat is being reported. A recent article on CNN.com quotes Dr. Debra Davis, Director of the University of Pittsburgh's Center for Environmental Oncology, saying that "You're just roasting your bone marrow" and asking "Do you really want to play Russian roulette with your head?" The article goes on to give five recommendations for limiting your exposure to cell phone radiation: Using a headset, using the speakerphone, getting a different phone, and so on. CNN followed up with another article with more quotes from Dr. Davis, this time saying that children are especially at risk because their brains are still developing, so they should be allowed to use cell phones in emergencies only.

As the director of an oncology center, she must have all kinds of experience treating cancer patients, and since she's going on CNN to talk about cell phone risks she must have a lot of experience dealing with cancer caused by cell phones. Right? Well, you'd think, but apparently CNN is not quite that particular about their guests. Dr. Davis' Ph.D. is in "science studies", whatever that is, and she is neither a medical doctor nor does she have any specialization in physical sciences like radiation. Now, I'm not trying to disrespect Dr. Davis — she has a fine background loaded with experience and all sorts of publications and accolades in her field — but I do want to draw attention to the fact that when CNN brings a doctor onto television to talk about a health problem, you shouldn't take anything for granted. You're the one who assumed that she treats cancer patients and has seen harmful effects from cell phone radiation. The fact is that the only danger Dr. Davis actually cited was that "since cell phones have only been in widespread use for 10 years or so, the long-term effects of their radiation waves on the brain has yet to be determined." Neither she, nor CNN, cited a single case of harm being caused by a cell phone, nor did they present any theoretical arguments indicating any plausible danger.

Dr. Davis is also dramatically wrong on one very significant point: That there has not yet been time for long-term studies to have been conducted, or that the question of cell phones and cancer is otherwise inadequately studied. In fact, the Journal of the National Cancer Institute published the results of a massive study in Denmark that followed the cancer histories of 420,000 cell phone users over 13 years. You'd think that someone in Dr. Davis' position would know about that, or at least take the slightest trouble to search for studies before going on CNN to proclaim that no such studies exist. The study's main interest was to search for increased incidences of brain or nervous system cancers, salivary gland cancer, and leukemia. The study concluded:

> *Risk for these cancers ... did not vary by duration of cellular telephone use, time since first subscription, age at first sub-*

scription, or type of cellular telephone (analogue or digital). Analysis of brain and nervous system tumors showed no statistically significant [standardized incidence ratios] for any subtype or anatomic location. The results of this investigation ... do not support the hypothesis of an association between use of these telephones and tumors of the brain or salivary gland, leukemia, or other cancers.

The lack of any connection is not surprising, given that no plausible hypothesis exists for how a cell phone could cause tissue damage. RF below the visible spectrum, which includes the frequencies used by cell phones and all radio devices, is not ionizing radiation and so has no potential to damage living cells or break any chemical bonds. Microwave ovens, which operate just above cell phones on the frequency scale, work by oscillating such an extremely powerful field back and forth, causing the water molecules to rub against each other and create heat by friction. Cell phone signals are three orders of magnitude weaker, too weak to move the water molecules, and do not oscillate to cause friction. Scratch the heat hypothesis, scratch the ionizing radiation hypothesis, and there are no plausible alternatives. Of course it's not possible to prove that there is no potential for harm, but all sources of harm known or theorized to date are clearly excluded.

So if that's true, how did the story get started? How did cell phones causing cancer become one of our pop culture myths?

It started in 1993, when a guy named David Raynard went on CNN's Larry King Live to talk about his lawsuit against the cellular phone industry over the death of his wife from brain cancer, who used a cell phone. Certainly we all sympathize with Mr. Raynard, but that doesn't make him right. Unfortunately for rationalism, being on Larry King was all the credibility the story needed to become a popular belief. Despite Mr. Raynard's claim that his wife's tumor was in the same shape as the cell phone antenna, the case was thrown out for a lack of evidence.

Another reason the belief persists is that it is constantly being promoted by companies selling quack devices claimed to protect consumers from any potential threat. Spreading fear is a major marketing angle that they employ. Cardo Systems, a maker of cell phone headset, broadly promoted as the best way to minimize danger of radiation, famously released a set of hoax videos on YouTube showing people popping popcorn by setting some kernels on a table between several activated cell phones. When nailed for the hoax by CNN, Cardo's CEO claimed that the videos were meant only as a joke and that the thought of scaring people into thinking that cell phones could pop popcorn never entered their minds. You can judge the credibility of that statement for yourself.

There are also a number of videos on YouTube showing eggs being hard-boiled merely by placing them between two activated cell phones for a few minutes. This claim has also been thoroughly debunked. The British TV show *Brainiac* even tried it with 100 phones. The result? Zippo. It didn't change the egg's temperature at all. Raw as ever.

Some of these companies selling products to protect you have sections on their web sites where they cite official statements reiterating that there is no proof that cell phones are safe. They also tend to cite one particular study, known as the Guy study and published in *Bioelectromagnetics* in 1992. The Guy study exposed rats to high levels of RF for 22 hours a day for two years. 18 of the exposed rats developed tumors, while only 5 of the control group did. The cell phone accessory companies stop there, but you have to dig deeper to find that other researchers have been unable to replicate these results, and the conclusion was that the tumor incidence, while statistically significant, was not shown to have been caused by the RF. In fact, another study also published in *Bioelectromagnetics* by Adey et. al. exposed rats to a chemical carcinogen and then exposed some of them to RF. Dr. Adey actually found *fewer* tumors in the RF exposed rats, but again the result was not large enough to draw conclusions. Even in the harshest of animal studies, no evidence has been found to link cell phone radiation to health problems.

We may quarrel with these companies' ethics in promoting fear to sell their products, but that doesn't mean that the products aren't a wise precaution. It can't hurt to be safe rather than sorry, can it? Well, you will be sorry if you spend any of your hard-earned money on a product intended to protect you from cell phone radiation, and you hear what the World Health Organization has to say on the matter. Their summary on such devices says:

> *Scientific evidence does not indicate any need for RF-absorbing covers or other "absorbing devices" on mobile phones. They cannot be justified on health grounds and the effectiveness of many such devices in reducing RF exposure is unproven.*

So far, the science that's been done pretty much supports the default skeptical position. When we hear a claim like "cell phone radiation causes cancer", we assume the null hypothesis until evidence is presented that supports the claim. And to date, all the good evidence supports the null hypothesis, not the claim. Maybe tomorrow things will change, and we'll find that cell phones are harmful, or that 60-cycle electrical outlets are harmful, or that traveling faster than 30 miles an hour is harmful. An open skeptical mind is open to any good evidence supporting any claim. But for now, I'm going to continue enjoying the usefulness of my iPhone, and be damn glad there's a tower in my neighborhood.

References & Further Reading

Cohen, E. "5 tips to limit your cell phone risk." *CNNHealth.com.* Cable News Network, 31 Jul. 2008. Web. 13 Jan. 2010. <http://www.cnn.com/2008/HEALTH/07/31/ep.cell.phones.cancer/index.html>

Johansen, C., Boice Jr., J., McLaughlin, J., Olsen, J. "Cellular Telephones and Cancer: A Nationwide Cohort Study in Denmark." *Journal of the National Cancer Institute.* 7 Feb. 2001, Vol 93, No 3: 203-207.

Muscat, J., Hinsvark, M., Malkin, M. "Mobile Telephones and Rates of Brain Cancer." *Neuroepidemiology.* 3 Jul. 2006, Vol 27, Issue 1: 55-56.

Shermer, M. "Can You Hear Me Now? The Truth about Cell Phones and Cancer: Physics shows that cell phones cannot cause cancer." *Scientific American.* 4 Oct. 2010, Volume 302, Number 10.

Tahvanainen, K., Niño, J., Halonen, P., Kuusela, T., Alanko, T., Laitinen, T., Länsimies, E., Hietanen, M., Lindholm, H. "Effects of cellular phone use on ear canal temperature measured by NTC thermistors." *Clinical Physiology & Functional Imaging.* 1 May 2007, Vol 27 No 3: 162-172.

WHO. "Electromagnetic fields and public health: mobile telephones and their base stations." *World Health Organization.* World Health Organization, 1 Jun. 2000. Web. 13 Jan. 2010. <http://www.who.int/mediacentre/factsheets/fs193/en/>

Wilson, J. "What to know before buying your kid a cell phone." *CNN.com Technology.* Cable News Network, 11 Aug. 2008. Web. 13 Jan. 2010. <http://www.cnn.com/2008/TECH/ptech/08/11/cellphones.kids/index.html>

3. It's Raining Frogs and Fish

Today we're going to run in panic from a meteorological downpour only Bartholomew and his Oobleck could appreciate: Storms of frogs and fish falling from the sky! For at least 200 years, newspapers and books have published accounts of people being pelted by huge numbers of frogs and fish coming down during rainstorms, or even sometimes out of a clear blue sky.

In 1901, a rainstorm in Minneapolis, MN produced frogs to a depth of several inches, so that travel was said to be impossible. Fish famously fell from the sky in Singapore in 1861, and again over a century later in Ipswich, Australia in 1989. Residents in southern Greece awoke one morning in 1981 to find that a shower of frogs had blanketed their village. Golfers in Bournemouth, England found herring all over their course after a light shower in 1948. In 1901, a huge rainstorm doused Tiller's Ferry, SC, and covered it with catfish as well as water, to the point that fish were found swimming between the rows of a cotton field. In 1953, Leicester, MA was hit with a downpour of frogs and toads of all sorts, even choking the rain gutters on the roofs of houses. The stories go on and on: More frogs in Missouri in 1873 and Sheffield, England in 1995, and more fish in Alabama in 1956.

How could such things happen? Obviously frogs and fish are heavier than air and can't evaporate up into clouds, nor can they suspend themselves up there to breed. Almost every printed version of these tales offers a single explanation: That a waterspout somewhere sucks the animals out of some water and lifts them up into the clouds, from where they later fall back to land. This explanation is so ubiquitous that even the Encyclopedia Britannica suggests it as the only offered hypothesis.

I've always had problems with the waterspout story, and the more you look into it, the poorer an explanation it turns out to be. Waterspouts come in two varieties, just like tornadoes on land. The first and most common is a non-tornadic waterspout, which is a local fair-weather phenomenon, akin to a dust devil you might see over farmland. They have little or no effect on the surface of the water. The second much rarer type is the full-blown supercell tornadic waterspout. The decreased air pressure inside a tornadic waterspout can actually raise the water level by as much as half a meter, but water itself is not sucked up inside. The visible column of a waterspout is made up of condensation, and is transparent. The high winds will kick up a lot of spray from wavelets on the surface, but if you look at pictures of waterspouts, you'll see that this spray is thrown outward, not sucked up inward. Just below the surface of the water, things are undisturbed. Waterspouts simply do not have any mechanism by which they might reach down into the water, collect objects, and then transport them upward into the sky.

If you've watched video of destructive tornadoes on land, you've seen this same effect. When a tornado rips through a building or a town, you'll see debris kicked up into the air, often quite high, from where it takes a ballistic trajectory outward. Never do objects ascend the inner column, because there is simply no mechanism inside for doing that. It's not an elevator; it's a destructive force scraping stuff off the surface and throwing it upward and outward. Stuff might take a lap or two around the column while it's being snatched up and tossed. Debris goes everywhere; groups of related objects are never picked up from one place, kept together, and neatly deposited somewhere else. Certainly there is no mechanism that might carry a group of objects way up into the clouds, transport them laterally great distances through fair weather while somehow counteracting gravity, and then suddenly release them in a single tight group to drop to the ground.

Not once in a single case of several dozen that I read was there ever a report of a tornado or waterspout in the vicinity, or even at all, no matter how far away. I conclude that waterspouts have no connection, either hypothetical or evidentiary, to the phenomenon of frogs, fish, or any other animals, falling out of the sky. There's a much better explanation that's well known to zoologists, but for reasons I can't fathom, is almost never put forward to explain these stories.

The thing is, we've got these stories repeated over and over again, and all of them, or almost all of them, are completely credulous. Nearly every author uncritically repeats the story, often giving the waterspout theory as a possible explanation. Almost never will you hear someone ask the question "Wait a minute; did this actually happen the way witnesses thought?" These authors don't know one of the fundamentals of critical examination: Before you try to explain a strange phenomenon, first see if the strange event ever really happened; or at least whether it happened the way it's been reported.

Drop a frog off a building, and unless it's extremely small, it goes splat; so undoubtedly, what people think they're seeing in these stories can't be what's actually going on. If the frog didn't come from the sky, could it have come from somewhere else?

Frogs do swarm naturally on occasion. It happens frequently enough that people start to correlate these events with other things that happened: Storms, earthquakes, celebrity deaths, what have you. It's been reported that frog swarms were correlated with both the 1989 Loma Prieta earthquake in California, and the 2008 Sichuan earthquake in China. Shortly thereafter, when frogs swarmed in Bakersfield, California, some called for earthquake preparedness. Needless to say, there was no earthquake; just a random population explosion of frogs. Sometimes these explosions can be dramatic. In 2004, four hurricanes hit Florida, making that state about the wettest it's ever been. The local species of frogs and toads all had a banner year, described by the Florida Museum of Natural History as a carpet.

Every spring and fall, frogs migrate between shallow breeding ponds and deeper lakes. Because they're amphibians and need to keep their skin moist, they migrate most often during rainstorms. In many cases, the day with the right conditions will come, and the whole frog population will move cross-country en masse, across roads, across properties, wherever it needs to go. When you look outside during a heavy rainstorm and you see thousands of frogs jumping everywhere all over the ground, the illusion that they're falling from the sky and bouncing can be quite convincing. A swarm of frogs looks like ping pong balls bouncing in a lottery machine. The fact that there usually aren't frogs here adds credibility to the illusion. Throw in a healthy dose of confirmation bias and some exaggerated second, third, and fourth hand reports, and you automatically end up with every imaginable detail like the frogs were choking rain gutters on top of buildings.

Although this explanation might satisfy the stories of frogs falling from the sky, what about fish? You don't find mass migrations of fish crossing overland, do you? Well, maybe not mass migrations, but believe it or not, there are fish species that occasionally take to the ground in search of better waters. There are many species of "walking fish" in the world. Mudskippers are probably the best known variety. In Florida in 2008, a school of about 30 walking catfish emerged from the sewer during a heavy storm and went slithering around on the street. The northern snakehead is another fish that can wriggle its way around on dry land. Throughout Africa and Asia are 36 species of climbing perches. They have a special organ that allows them to breathe air, and are able to walk using their gill plates, fins and tail, despite looking completely fishlike with no obvious ambulatory limbs. None of these fish move gracefully or even look like they have the ability. To the average witness, it's a live fish flopping around on the ground where no fish has any business being, and having fallen from the sky seems as good an explanation as any.

But they can't have fallen from the sky. Drop a fish off a building, and that's a dead fish. These fish are not reported as being burst open with their guts splattered out, but as flapping and squirming about, very much alive. An alternate explanation, that these fish are simply using their rare but well established ability to move overland, doesn't require us to accept some unexplained, implausible hypothesis even in the face of a lack of splattered-fish evidence.

We have no reason to think anyone actually observed the fish falling from the sky. All we know is that some people have reported finding live fish on dry ground. No doubt many of them couldn't think of any explanation other than the fish fell there, so they probably told the reporter "A bunch of fish fell out of the sky." At that point, the reporter had the story he was looking for, and needed to inquire no further.

Go back and read any story you've ever seen about frogs and fish falling from the sky, this time allowing for the possibility that the animals were already naturally on the ground when the witnesses first discovered them. Allow for the possibility that some elements of the report, like the falling part, could be based originally on witness conjecture or assumption. What you'll find is that a tale so perplexing that the only possible explanation seemed to be a paranormal event, has now become a very cool example of rare animal behavior that most people don't even know about. When you fail to think critically, you fail to learn new information. Never stop your investigation prematurely at the popular supernatural explanation, and always be skeptical.

References & Further Reading

Camero, H. "A Big Night for viewing frogs and salamanders." *Wicked Local Bolton.* GateHouse Media, Inc., 28 Mar. 2008. Web. 26 Jan. 2010.
<http://www.wickedlocal.com/bolton/homepage/x1681298480>

Fort, Charles. *The Book of the Damned: The Collected Works of Charles Fort.* New York: Jeremy P. Tarcher/Penguin, 2008. 42-50, 81-99, 299-305.

Graham, Jeffrey B. *Air-breathing fishes: evolution, diversity, and adaptation.* San Diego: Academic Press, 1997. 54-56.

Gudger, E.W. "Rains of Fishes." *Natural History.* Natural History Magazine, 1 Nov. 1921. Web. 8 Sep. 2009.
<http://www.naturalhistorymag.com/picks-from-the-past/271577/rains-of-fishes>

Schneider, T. "The Vernal Pool A Place of Wonder." *Wild Ones Journal.* 1 Mar. 2006, Volume 18, Number 2.

4. The Battle of Los Angeles

Today we're going to turn the pages back to an American UFO story dating from World War II, the Battle of Los Angeles, when (according to modern lore) the United States Army and Navy battled a giant UFO hovering above the city of Los Angeles.

It was late February, 1942, less than three months after the Japanese attacked Pearl Harbor. Residents on the Western coast of the United States expected they were next, and so stood ready with hasty fortifications and kept their eyes on the sky. The crews manning the antiaircraft artillery batteries in Los Angeles had been trained, but lacked experience in actual combat. Only one day before, the Japanese submarine I-17 had surfaced off of Santa Barbara and fired 25 shells at some aviation fuel storage tanks, so the alert level was the highest it had ever been. An attack on Los Angeles was imminent.

Just after 2:00am on the morning of February 25, radar picked up a target off the coast. The antiaircraft batteries in Los Angeles were put on Green Alert, ready to fire. By 2:21am the radar target had approached closer, and a blackout was ordered. The radar lost contact with its target, and searchlight beams swept the sky for nearly half an hour. Then, reports of aircraft came in. Over Santa Monica, a balloon carrying a red flare was spotted, and the batteries opened fire at 3:06am. The Battle of Los Angeles was on.

For nearly an hour, batteries fired 1,430 rounds of antiaircraft artillery, raining eight and a half tons of shrapnel back down onto Los Angeles. But what did they see? What were they shooting at? Therein lies the rub. Many saw nothing. Some reported balloons. A few reported airplanes. CBS Radio called it a blimp. The moon had set at 2:30am, and sunrise was not until 6:30am; combined with the blackout, it was about as dark as dark can be. The only thing anyone could see was whatever the searchlights struck, which was smoke from the AAA bursts. The Office of Air Force History described the field reports as "hopelessly at variance". The most famous photograph, from the Los Angeles Times, shows a convergence of searchlights onto a single large cloud of smoke. Property damage from the shrapnel was widespread, and since no bombs were dropped and no evidence of enemy aircraft was ever discovered, demands for explanations and investigations followed: Both in a scathing editorial in the Los Angeles Times the following day, and from the White House.

Secretary of the Navy, Frank Knox, held a press conference that same day to state that it was a false alarm, that no aircraft had been involved, and that the entire incident had been an expensive case of jittery nerves. Chief of Staff George Marshall wrote a memo to President Roosevelt, stating the current understanding that airplanes may or may not have been involved, possibly as many as fifteen, possibly commercial aircraft, at various slow speeds. Given the lack of confirmation that any aircraft were present at all, Roosevelt's response was to ask the Secretary of War to clarify exactly who is authorized to order an air alarm.

And that's where the story was left for decades: a false alarm from the opening days of World War II: No mysteries, no strangeness, no aliens, no supernatural element. But of course, as you can guess, it did all eventually appear. It took more than 40 years, but UFO enthusiasts finally decorated the Battle of Los Angeles with some imaginative additions.

To understand how it happened, you first have to understand the Majestic 12 papers. In 1987, a group of UFOlogists, William L. Moore, Stanton Friedman, and Jaime Shandera, announced the existence of several government documents, classified as top secret, that purported to contain a 1947 order from President Harry Truman establishing a group called Majestic 12, an assortment of the usual Illuminati from government, business, and the military. Majestic 12 was charged with handling everything to do with extraterrestrial aliens.

Later, another UFOlogist, Tim Cooper, announced his own batch of secret Majestic 12 documents. Rival UFOlogists work together in the same way that rival Bigfoot hunters do: Not very nicely. Moore and his proponents launched into Cooper's documents, pointing out clues that prove them counterfeit; and Cooper and his proponents did the same to Moore's documents, revealing the flaws that disproved their authenticity. When infighting among adversarial bamboozlers does all the work revealing each others' hoaxes, it makes the legitimate investigator's job so much easier.

Among this tangled mess of hoax documents is a letter called the Marshall/Roosevelt Memo from March 5, 1942, stating that two unidentified aircraft were in fact recovered after the Battle of Los Angeles: One at sea, and one in the San Bernardino Mountains east of Los Angeles. It says in part:

> *This Headquarters has come to a determination that the mystery airplanes are in fact not earthly and according to secret intelligence sources they are in all probability of interplanetary origin.*

The letter is, of course, properly scuffed up and smudged in the most realistic and dramatic fashion. A PDF of it is available for download from MajesticDocuments.com. Hilariously, page 2 of the PDF is an order form to purchase a wide range of UFO related documents, CDs, and books. Obviously, it's not legal to distribute actual top secret documents, and the fact that the FBI permits the availability of this (and the many others on MajesticDocuments.com) is a pretty good tipoff to the FBI's assessment of their authenticity. Skeptical investigator Philip Klass brought the documents' publication to the FBI's attention in 1988, and the FBI quickly concluded that all the documents were fake. So download freely, and send in those order forms.

As far as I could determine, this letter's late-1980's appearance was the earliest reference to anything UFO related happening at the Battle of Los Angeles. Since then, of course, innumerable references have appeared on the web. Most UFO web sites discuss the battle and show the picture from the LA Times, describing the cloud of AAA smoke in the searchlights as a "large craft". But this was not the contemporary identification. For more than 40 years, not a single person associated with the Battle of Los Angeles entertained any thoughts about extraterrestrial spacecraft or aliens, according to all available evidence (at least when you discard the hoaxed evidence). The alien spacecraft angle is purely a post-hoc invention by modern promoters of UFO mythology.

Modern UFOlogists seem to have forgotten what the "U" in UFO stands for: Unidentified. They tend to identify such objects as extraterrestrial spacecraft, for reasons known only to themselves; so they should really pick a new term. The Battle of Los Angeles was triggered by true UFO's: Something spotted in the sky that nobody was able to definitively identify. Most gunners reported never seeing anything at all, and simply fired at wherever they saw other air bursts. For this, the gun crews were officially reprimanded. The Office of Air Force History says in its 1983 report entitled *The Army Air Forces in World War II:*

> *A careful study of the evidence suggests that meteorological balloons — known to have been released over Los Angeles — may well have caused the initial alarm. This theory is supported by the fact that anti-aircraft artillery units were officially criticized for having wasted ammunition on targets which moved too slowly to have been airplanes. After the firing started, careful observation was difficult because of drifting smoke from shell bursts. The acting commander of the anti-aircraft artillery brigade in the area testified that he had first been convinced that he had seen fifteen planes in the air, but had quickly decided that he was seeing smoke. Competent correspondents like Ernie Pyle and Bill Henry witnessed the shooting and wrote that they were never able to make out an airplane.*

But of course, to the conspiracy theorists and UFO believers, any report put forth by the Air Force is simply part of the conspiracy and not to be trusted. So let's play the devil's advocate and assume that interplanetary spacecraft were, in fact, shot down during the battle and recovered, and the government has full knowledge of it, as the UFOlogists expect us to believe. Then it becomes a question of how they were able to keep this a secret for more than 40 years: Retroactively change the newspaper accounts, change the radio reports, pay off or kill everyone who participated, pay off or kill everyone in Los Angeles who witnessed it, yet continue to allow the "top secret" confessions to be downloadable from the Internet; the proposition quickly becomes ludicrous.

An alternate explanation, supported by evidence, requires us to make no such absurd leaps of logic or pseudoscientific assumptions: That the Battle of Los Angeles was simply a case of jittery nerves, at a time when every single person in Los Angeles was living in daily fear for their lives from imminent Japanese attack. There is simply no need for the introduction of a paranormal element to explain it. Whenever you hear a tale from history that involves alien spacecraft or any other paranormal element, you should always be skeptical.

REFERENCES & FURTHER READING

Craven, W., Cate, J. *The Army Air Forces in World War II, Vol. 1.* Washington, D.C.: Office of Air Force History, 1983. 277-286.

FBI. "MAJESTIC 12." *MAJESTIC 12.* Federal Bureau of Investigation, 28 Aug. 1991. Web. 15 Sep. 2009. <http://foia.fbi.gov/foiaindex/majestic.htm>

Friedman, Stanton T. *Top Secret/Majic.* New York: Marlowe & Co., 1996.

Klass, Philip. "The New Bogus Majestic 12 Documents." *Skeptical Inquirer.* 1 May 2000, Volume 24, Number 3.

Knight, Peter. *Conspiracy Theories in American History: An Encyclopedia.* Santa Barbara, CA: ABC-CLIO, 2003. 700.

5. Demystifying the Bell Witch

Along Highway 41 in the hamlet of Adams, Tennessee, amid green fields and trees, stands Tennessee Historical Marker 3C38, entitled The Bell Witch. The solitary marker tells the following tale:

> *To the north was the farm of John Bell, an early prominent settler from North Carolina. According to legend, his family was harried during the early 19th century by the famous Bell Witch. She kept the household in turmoil, assaulted Bell, and drove off Betsy Bell's suitor. Even Andrew Jackson, who came to investigate, retreated to Nashville after his coach wheels stopped mysteriously. Many visitors to the house saw the furniture crash about them and heard her shriek, sing, and curse.*

The Bell Witch story is frequently promoted with two popular claims: That it's the only haunting known to have actually killed a person, and that it's the only haunting to directly involve a US President. Let's briefly summarize the legend.

In 1817, John Bell encountered a strange animal in his field: It had the body of a dog, but the head of a rabbit. For some time thereafter, the Bell family was tormented by pounding on the outside of their farmhouse every night. They would rush out hoping to catch the strange animal, but never found anything. The noises moved indoors — scratching and slamming and strange whispers made sleep nearly impossible. Sometimes pillows and blankets were whisked away by an unseen force. The Bells' youngest daughter Elizabeth, nicknamed

Betsy, got the worst of it. She was often slapped and had her hair pulled. Friends who spent the night with the Bells to help were subjected to the same torments. The whispers grew louder and became a disembodied female voice, singing hymns and quoting scriptures, and carried on conversations with the Bells and their guests. Word of the disturbances spread and in 1819 reached Andrew Jackson, a heroic Major in the US Army from the Battle of New Orleans.

Jackson and his men stayed at the Bell homestead to investigate. One of Jackson's men was physically attacked and beaten by an unseen force, and when Jackson himself finally gave up and fled the farm he said "I'd rather fight the entire British Army than to deal with the Bell Witch." Ten years later, he became our 7th President.

The witch's wrath focused increasingly upon John Bell himself, driving him into frail health, until one night in 1820, he was found on his deathbed with a vial of a strange potion. To see what it was, the family gave some of it to the cat, which immediately dropped dead. The witch laughed and sang, and boasted "I gave Ol' Jack a big dose of that last night, and that fixed him." John Bell was dead, but the hauntings continued, and the legend lived on.

If the name of the Bell Witch sounds familiar, it may be the similarity to the 1999 movie *The Blair Witch Project*, which is said to have been at least inspired in part by the Bell Witch story. Following the success of *The Blair Witch Project*, a rash of movies about the Bell Witch came out hoping to capitalize on the popularity: *Bell Witch Haunting* in 2004, *An American Haunting* in 2005, *Bell Witch: The Movie in 2007*, and IMDB lists a movie called simply *Bell Witch* as being in production.

Now a lot of you are probably saying something like "Oh, another old ghost story, whoop-de-do, Dunning, I wonder what your opinion is going to be on that." Well, I've been doing the science journalism thing a long time, and if there's one thing I've learned, it's that every story, no matter how familiar or seemingly obvious, presents some new challenge in honing your critical thinking skills. We've talked about two other popular hauntings on the Skeptoid podcast — Borley Rectory and The Amityville Horror — and both turned out to be fabrications by authors. You can't simply ask "Gee, the Amityville family found evil cloven footprints in the snow, how do you explain that?" because the question is based on a false presumption: That such an event actually took place. To use the scientific method to uncover the truth about the Bell Witch, you can't take anything for granted; and before you take the trouble to examine the specific claims, you need to look at the source of the claims to see if there's actually anything to examine.

We start by looking at the published accounts. There have been so many books written about the Bell Witch that I'm not even going to bother naming them. But, for their sources, they all draw upon the earliest book, *Authenticated History of the Bell Witch* from 1894, by Martin Van Buren Ingram, owner of a regional newspaper. This was the first book published about the Bell Witch, and it was published *75 years* after the hauntings. That's a long time. Long enough that the author wasn't even born when the hauntings took place. So what was his source?

Martin Ingram's book is based entirely upon the handwritten diary of Richard Bell. Richard Bell, one of John Bell's sons, was born in 1811, so he was about six years old when the hauntings began. According to Ingram, Richard waited until 1846, more than 30 years, before he actually wrote down the events in his diary. He recorded his 30 year old memories of being a six year old child. Ingram goes on to say that in 1857 Richard gave the diary to his son, Allen Bell, who subsequently (and quite inexplicably) gave it to Ingram, with instructions to keep it private until after the deaths of the immediate family. That happened around 1880, when Ingram began writing his book. Conveniently, every person with firsthand knowledge of the Bell Witch hauntings was already dead when Ingram started his book; in fact, every person with *secondhand* knowledge was even dead.

Martin Ingram never said anything about what became of this alleged diary. There is no record of anyone else having seen it, and logically, Ingram should have promoted the diary's existence in his newspaper to publicize his book. He did not. I am certainly not convinced that the diary ever existed at all. Why would Richard Bell wait 30 years to write down such an incredible story? Why would Allen Bell give away such a unique heirloom to Ingram? Those are big questions, and Ingram had every reason to falsify the diary's existence.

Ingram's book also falsified at least one other source. His book claims that in 1849, the *Saturday Evening Post* ran a story about the Bell Witch, blaming the crazy daughter Elizabeth for everything, and then retracted the story shortly thereafter once she threatened to sue. People have looked for such an article and none was ever found. I called the *Saturday Evening Post*, and was told that their microfilmed archives for that period no longer exist. Researcher Jack Cook went through other microfilms of the *Post* for several years on either side of 1849 and confirmed that no such article was ever published. Even people looking for it in 1894, following the publication of Ingram's book, failed to find such an article; which casts doubt on Ingram's own ability to have found it. Without exception, all of Ingram's sources for his book were conveniently untraceable.

Historians have found only one printed reference to the Bell Witch that predates the publication of Ingram's book, and it's a brief one-paragraph blurb in the 1886 first edition of *Goodspeed's History of Tennessee* in its chapter on Robertson County, which reads as follows:

> *A remarkable occurrence, which attracted wide-spread interest, was connected with the family of John Bell, who settled near what is now Adams Station about 1804. So great was the excitement that people came from hundreds of miles around to witness the manifestations of what was popularly known as the "Bell Witch." This witch was supposed to be*

> some spiritual being having the voice and attributes of a woman. It was invisible to the eye, yet it would hold conversation and even shake hands with certain individuals. The freaks if performed were wonderful and seemingly designed to annoy the family. It would take the sugar from the bowls, spill the milk, take the quilts from the beds, slap and pinch the children, and then laugh at the discomfiture of its victims. At first it was supposed to be a good spirit, but its subsequent acts, together with the curses with which it supplemented its remarks, proved the contrary.

Notice the two most significant events are missing: The witch's murder of John Bell, and Andrew Jackson's involvement. No newspapers described either event. No court records or recorded minutes from churches described either event. The story of John Bell's murder at the hands of the Bell Witch was never described in any published account, nor placed into the pop culture version of events by the frightened family's reports. It seems almost incredible ...unless Ingram made it up.

Ingram almost certainly made up the entire Andrew Jackson incident. Andrew Jackson's whereabouts between 1814 and 1820 are well documented, and there is no known record of his having visited Robertson County during those years. In all of his own writings and in all of his many biographies, there is not a single mention of his alleged Bell Witch adventure. The 1824 Presidential election was notoriously malicious, and it seems hard to believe that his opponent would have overlooked the opportunity to drag him through the mud for having lost a fight to a witch. All known documentation shows Jackson elsewhere during the period in question, and all published material about his encounter with the Bell Witch relies on Martin Ingram's book as its sole source.

So what evidence of the Bell Witch are we left with? Vague stories that there was a witch in the area. All the significant facts of the story have been falsified, the others come from a source of dubious credibility. Since no reliable documentation of any actual events exists, there is nothing worth looking into. Ingram also wrote that the Bell Witch promised to return in

1935, and since nothing happened in that year either, I chalk up the Bell Witch as nothing more than one of many unsubstantiated folk legends, vastly embellished and popularized by an opportunistic author of historical fiction.

References & Further Reading

Cook, Jack. "The Spirit of Red River." *Bell Witch Legend.* Jack Cook, 1 Sep. 2006. Web. 16 Jan. 2010. <http://bellwitchlegend.blogspot.com/>

Goodspeed Brothers. *History of Tennessee.* Nashville: Goodspeed Publishing Co., 1886.

Hudson, A., McCarter, P. "The Bell Witch of Tennessee and Mississippi: A Folk Legend." *American Folklore Society.* 31 Mar. 1934, Volume 7, Number 183: 45-63.

Ingram, Martin Van Buren. *An Authenticated History of the Bell Witch.* Clarksville: M. V. Ingram & Co., 1894.

Middle Tennessee Skeptics. "Bell Witch." *Middle Tennessee Skeptics.* Middle Tennessee Skeptics, 21 Aug. 2008. Web. 14 Feb. 2010. <http://mtskeptics.homestead.com/>

6. Falling Into Mel's Hole

Today we're going to point our skeptical eye downward, down into the deepest hole in the world. Somewhere in the hills of eastern Washington state is said to be a bottomless pit. Not only can you throw as much junk as you want into the hole without it ever hitting bottom or the hole ever filling up, it has other stranger properties. A black beam sometimes shoots up out of the pit, like a solid shadow. A beloved pet dog that had died was once disposed of in the pit, only to come trotting happily out of the woods hours later, very much alive again. Radios brought near the hole play old-time radio broadcasts. The place has some kind of weird aura that makes animals avoid it.

Mel's Hole offers a rare opportunity for seekers of the unknown, because a hole in the ground is something physical that doesn't move and that's always going to be there if you want to go and see it and test it. So I found the idea intriguing, and trust me, if there was any indication something like Mel's Hole existed, I'd make every effort to go check it out. But I quickly hit a snag: Apparently, there is no such place. Mel's Hole appears to be nothing more than the pipe dream of a series of crank calls into the Coast to Coast AM radio show, beginning in 1997.

A guy who said his name was Mel Waters made five calls into Coast to Coast AM between 1997 and 2002, and told his story to the host, Art Bell. Mel said he'd bought the property in Washington and was aware that all the locals knew about the strange hole, and that everyone routinely used it for garbage disposal. Old refrigerators, used tires, even dead cattle were tossed in all the time, and yet the hole never filled up. Physically the hole looked like a well, about nine feet across, walled with stone, and with a 3-4 foot high stone wall surrounding it. Personally I'd like to have seen the operation that tiled the walls of a bottomless pit with stone, even if they only did the top portion. Must have been quite the acrobatic performance. Mel Waters said that 20 people used the hole regularly for disposal.

In his latter Coast to Coast AM appearances, Mel stated that government agents in yellow suits raided his property, seized it, and paid him $250,000 a year to move to Australia and never discuss the hole again. They also removed it from Google Earth and doctored whatever public records were necessary to eliminate any evidence that Mel, his property, or his hole, had ever existed. In logic we call this a special pleading: An excuse invented to explain away any objection. Why is there no evidence? Of course, the "government" eliminated all the evidence.

A gentleman named Gerald Osborne, who inexplicably calls himself Red Elk despite claiming no Native American heritage, was the next to call into Coast to Coast AM to discuss Mel's Hole, in September of 2008. Red Elk carries a piece of what he believes to be an alien spacecraft on a necklace, and lectures about impending apocalypse and Reptoids who live beneath the surface of the Earth. Red Elk states that the hole is between 24 and 28 miles deep, though he declines to explain how those limits were established. Red Elk has variously said either that he visited Mel's Hole as a young boy with his father in 1961, or that he "goes there quite often." Yet, in a 2002 *Seattle Times* article documenting an expedition of enthusiasts trying to find Mel's Hole, Red Elk himself was there to lead the party, and yet he didn't seem to have any more idea where to look than anyone else did.

Mel stated that he'd tried to measure the depth of the hole by lowering a weighted fishing line into it. He told Art Bell he'd lowered 80,000 feet of fishing line into the hole with a one-pound weight at the end. He didn't say what pound test the line was, or whether it was monofilament or braided. He did say he bought it in 5000-yard spools. I found these available online for $40 for the thinnest, cheapest 10-pound test line all the way up to $1300 for braided, hollow core 130-pound line. So if Mel's on his sixth spool, he's got anywhere between 48 and 230 pounds of fishing line hanging, according to the published shipping weights of those spools that I was able to find. Would it hold?

You also have to consider the temperature at that depth. Since we don't know where Mel's Hole is supposed to be, we don't know the thickness of the Earth's crust there. Washington state runs 65,000 to 130,000 feet thick. So, it's possible that Mel's 80,000 feet is through solid rock and not viscous magma. We can calculate that the temperature of the rock at that depth is probably around 1300 degrees Fahrenheit, or about 715 degrees Celsius; way hotter than any fishing line can exist. But who knows, perhaps the unprecedented ability to remain cool and solid at any depth is another of Mel's Hole's supernatural characteristics. That would be another special pleading.

Mel himself seems to be something of a mystery. Enthusiasts of the hole have searched public records extensively, and found nobody of that name living or voting or paying taxes in the area, found no unaccounted private property in the area, and found no property transfers that fit Mel's timeline of when he said he acquired the land. In short, every detail that Mel did share with Coast to Coast AM that was falsifiable, has been falsified by amateur investigators interested in finding the hole. We have to conclude that "Mel Waters" was almost certainly a fake name.

It wouldn't be the first or the last time Coast to Coast AM got hoaxed. In January 2008, the show received a call from a guy calling himself Gordon, a theoretical physicist who worked at a research facility. Gordon described some strange events that were going on at his workplace, and their work involving "portal technology", which host George Noory embraced uncritically. Video gamers recognized the story at once. The caller was pretending to be Gordon Freeman, the character from the video game Half-Life. Perhaps inspired by this, someone else called the show in November 2008 to relate his dream, which was the story from the game Fallout 3. Now it's certainly not fair to blame George Noory or the show's producers for being unfamiliar with these games; you could probably describe the plots of most video games to me and I'd have no idea what you're talking about either. But you can fairly criticize the show for having no meaningful screening process. These episodes show that Coast to Coast AM will gladly give a platform to any caller, with any story, no matter how preposterous, even those made up on the spur of the moment with no documentation or verification. Now, there's nothing wrong with doing that and I don't think it's a fair criticism; after all it's a show about weird stuff and they make no representation that everything (or anything) they broadcast is true. So the lesson to learn is that there's no reason to think that stories promoted by Coast to Coast AM have been documented or independently verified. Gordon Freeman wasn't, and Mel's Hole certainly wasn't either.

But why did Mel's Hole strike such a nerve and convince so many people that it was real? Believers and the merely curious set up a web site, MelsHole.com, to share information and track progress on finding either Mel or his mysterious hole. The site generated more than 8,000 posts on nearly 600 topics until its moderator threw in the towel and acknowledged that no progress had been made in ten years since Mel called the Coast to Coast AM program, and that the site was effectively dead unless Mel came forward.

Mel's Hole is a pretty cut and dried case. There's not really anything there to interest a skeptical investigator. All we have is an anonymous person who called a radio show using a fake name and told an implausible, unverifiable story. All the people who *would* be able to back it up — the local ranchers and folks who had been dumping trash and cattle into it for decades — don't seem to exist either. If I dumped my trash into such an extraordinary hole, I'd bring friends back to see it. They'd bring friends. News of such an incredible hole would spread like wildfire. Is it plausible that it could have had such regular local use, and yet nobody would be aware of its existence? Who are these people? Why can't they be found? Surely the government has not relocated everyone in the county to Australia; there are no vacant homes and no fenced-off government facilities.

Never assume that implausible stories must be true simply because you're unable to disprove them. You never will be able to, because special pleadings can always be invented to explain away any questions you might raise. What *can't* be invented from thin air is verifiable evidence, and its absence in the case of Mel's Hole speaks loud and clear.

REFERENCES & FURTHER READING

Davis, J., Eufrasio, A., Moran, M. *Weird Washington: Your Travel Guide to Washington's Local Legends and Best Kept Secrets.* New York: Sterling Publishing Co. Inc., 2008. 50-52.

Harvey, D., et al. *Aspects of Mel's Hole: Artists Respond to a Paranormal Land Event Occurring in Radiospace.* Santa Ana: Grand Central Art Center, 2008.

Konen, K., Nurse, M. "Mel's Hole." *UWTV.* UW Independent Filmmaking, 1 Aug. 2004. Web. 2 Jun. 2009.
<http://www.uwtv.org/programs/displayevent.aspx?rID=5160&fID=1474>

Waters, M., Bell., A. "Transcripts of Mel Waters of 'Mel's Hole' Fame." *Mel Hole Transcripts.* The Seattle Museum of the Mysteries, 21 Feb. 1997. Web. 2 Jun. 2009.
<http://www.seattlechatclub.org/Mel_Hole_Transcripts.html>

Zebrowski, John. "Expedition seeks paranormal pit." *Seattle Times.* 14 Apr. 2002: B1.

7. The Truth about Aspartame

As a kid I remember hearing that the artificial sweetener aspartame would neutralize your digestive enzymes, and anything else you ate that day would turn to fat. Although this makes no sense biochemically at any level, it sounded scientific enough to me and I was satisfied with it — at least enough to justify my dislike for its awful flavor. It turns out that my fat-producing claim was about the mildest of many arguments made by a growing anti-aspartame movement, and biochemically-nonsensical as it was, it was among the sanest of the arguments. Take a look at websites such as AspartameKills.com, Sweet-Poison.com, and Dark-Truth.org, and you'll see that a whole new breed of aspartame opponents has taken activism to a whole new level.  Here are a few quotes from those web sites:

> Thank you Montel Williams for having the fortitude to say: "Multiple Sclerosis is often misdiagnosed, and that it could be aspartame poisoning"

> NutraSweet® killed my mother and has killed and/or wounded millions of innocent people in the US and abroad.

> Aspartame converts to formaldehyde in vivo in the bodies of laboratory rats.

> ARTIFICIAL SWEETENER, ASPARTAME, (EQUAL, NUTRASWEET) LINKED TO BREAST CANCER AND GULF WAR SYNDROME.

> *Did O J Simpson Have a Reaction to Aspartame that led to the deaths of Nicole Simpson and Ron Goldman?*
>
> *THE FDA, THE INTERNATIONAL FOOD INFORMATION COUNCIL (IFIC), PUBLIC VOICE AND OTHERS ARE SCAM NON-PROFIT ORGANIZATIONS AND PAWNS OF THE NUTRA-SWEET COMPANY.*
>
> *After more than twenty years of aspartame use, the number of its victims is rapidly piling up, and people are figuring out for themselves that aspartame is at the root of their health problems. Patients are teaching their doctors about this nutritional peril, and they are healing themselves with little to no support from traditional medicine.*
>
> *Donald Rumsfeld disregarded safety issues and used his political muscle to get Aspartame approved.*
>
> *The Nazi Scientist's Poison Projects: Poison adults with Aspartame*
>
> *Because of your and/or your forbearers [sic] exposure to toxics like Aspartame, a summation of immune, mitochondrial, DNA, and MtDNA (genetic) damage has occurred in your body that has made your body unable to deal with chemical insults.*
>
> *The doctor that was in charge of the lab to study Aspartame, reported that the substance was too toxic and he mysteriously dissappeared [sic] and all the paper work somehow was destroyed.*
>
> *The Nazis actually won the war. They just pretended to lose so that we wouldn't notice them take over our government.*

Well, that's enough for now. And if you haven't heard those, you've almost certainly received one of several hoax emails that people have been forwarding around since 1995, according to Snopes.com, giving an equally long list of untrue claims about aspartame being the cause of nearly every illness. One is even falsely attributed to Dr. Dean Edell. Suffice it to say that every possible kind of attack is made against aspartame: Pseudoscientific attacks where they throw out whole dictionaries of scientific sounding nonsense; guilt by association attacks where they mention aspartame alongside Adolf Hitler and Donald Rumsfeld; non-sequiturs like pointing out the evils of the corporate structure of pharmaceutical companies as if that is support for how and why an "aspartame detoxification" program will "heal" you of all disease; and even Bible quotations attacking aspartame. The anti-aspartame lobby appears to include everyone from alternative treatment vendors trying to sell their products, to fully delusional conspiracy theorists. Dr. Russell Blaylock, a retired surgeon turned anti-pharmaceutical author and activist, believes aspartame is part of a massive government mind-control plot:

> *We're developing a society because of all these different toxins known to affect brain function. We're seeing a society that not only has a lot more people of lower IQ, but a lot fewer people of higher IQ. In other words a dumbing down, a chemical dumbing down, of society. ...That leaves them dependent on government because they can't excel. ...So, you know, you can kind of piece it together as to why they are so insistent on spending so many hundreds of millions of dollars of propaganda money to dumb down society.*

Discovered in 1965 at Searle (now Pfizer), aspartame is an artificial sweetener, aspartyl-phenylalanine-1-methyl ester. Chemistry types call it a methyl ester of the dipeptide of the amino acids aspartic acid and phenylalanine. It is 180 times as sweet as sugar, which is why it's such an effective low-calorie sweetener: It's needed in only miniscule amounts. Partly in response to all the anti-aspartame craziness out there, a group of scientists from the NutraSweet company published a 2002 review of dozens of studies and clinical trials performed worldwide in the journal Regulatory Toxicology and Pharmacology, which made the following conclusions:

> *Over 20 years have elapsed since aspartame was approved by regulatory agencies as a sweetener and flavor enhancer. The safety of aspartame and its metabolic constituents was established through extensive toxicology studies in laboratory animals, using much greater doses than people could possibly con-*

sume. ...*Several scientific issues continued to be raised after approval, largely as a concern for theoretical toxicity from its metabolic components — the amino acids, aspartate and phenylalanine, and methanol — even though dietary exposure to these components is much greater than from aspartame. Nonetheless, additional research, including evaluations of possible associations between aspartame and headaches, seizures, behavior, cognition, and mood as well as allergic-type reactions and use by potentially sensitive subpopulations, has continued after approval. ...The safety testing of aspartame has gone well beyond that required to evaluate the safety of a food additive. When all the research on aspartame is examined as a whole, it is clear that aspartame is safe, and there are no unresolved questions regarding its safety under conditions of intended use.*

And yet the claims persist unabated. Here are a few more, addressed point-by-point:

- Claims that aspartame causes Multiple Sclerosis are entirely made up and have no evidence or plausible foundation. Search the Multiple Sclerosis Foundation's web site for "aspartame" to find more than enough information refuting this harmfully misleading claim.

- The idea that aspartame causes "methanol toxicity" is based on the fact that when digested, aspartame does release a tiny amount of methanol. It's less than the amount you get from eating a piece of most any fruit. Tomato juice, for example, gives you four times the methanol of a can of diet soda. It's a common, naturally occuring environmental compound that is found in many foods. Nancy Markle, one of the most vocal aspartame conspiracy theorists, charges that the autoimmune disease lupus is actually misdiagnosed methanol toxicity caused by drinking 3-4 cans of diet soft drinks per day. If she's right, everyone who drinks a glass of tomato juice each day (or the equivalent in other fruits) is gravely ill with lupus. *Time* magazine once devoted an entire article to debunking Nancy Markle's baseless claims about aspartame.

- Much has been made of the claim that aspartame turns into formaldehyde in your system. This is true, because formaldehyde is a natural byproduct of digestion of methanol, and it happens whenever you eat almost anything. Formaldehyde is carcinogenic and is considered very dangerous in cases of occupational exposure, for example, when you get a dosage many orders of magnitude greater than the trace amounts produced during natural digestion. Again, aspartame does this in much smaller amounts than many common foods, so this has been a normal, healthy component of digestion for as long as humans have been eating fruits and vegetables.

- Gulf War Syndrome is a weakly evidenced correlation between service in the Gulf War and incidences of chronic fatigue, chronic pain, and a range of vague neurological conditions. Anti-aspartame advocates blame aspartame for this, but there is no correlation between increased aspartame consumption and Gulf War service. In addition, aspartame is among the hypothesized causes that have been eliminated by the Research Advisory Committee on Gulf War Veterans' Illnesses.

- What about Donald Rumsfeld's involvement with aspartame? He was the CEO of Searle at the time aspartame was approved as a sweetener. The reason he was hired was as a financial turnaround wizard, which he accomplished; he wasn't the guy in the lab designing artificial sweeteners. Even if you accept the conspiracy charges that he leveraged his cronies to force approval of a potentially dangerous product, that still says nothing about aspartame. It's a giant non-sequitur. To assess the safety of a product, we don't ask "Who was the CEO and who were his cronies?" we ask "What are the test results?" and so far, all of the test results show no association between aspartame and any of the diseases it is claimed to cause.

- Some of the claims about aspartame breaking down into unwanted compounds are true at extreme levels of alkalinity or acidity, levels which would be fatal to a human being anyway. This is the case with many foods, by no means is this unique to aspartame. So if your body chemistry is such that aspartame would be a danger to you, you'd have to already be dead from some unrelated cause.

- There is actually one known health risk associated with aspartame, but it only applies to people with a rare genetic disorder on chromosome 12 called phenylketonuria or PKU, which affects about 1 in 15,000 people. They can't metabolize phenylalanine, so they need to minimize not only aspartame but all phenylalanine products. Phenylalanine is an amino acid that's found in many, many foods, including breast milk; so it hardly makes sense to single out aspartame as the problem product.

- It is true that you can't cook with aspartame, but not for any safety reasons. At cooking temperatures it breaks down and loses its flavor, like some other foods. For this reason aspartame is starting to lose ground in the market to sucralose, another artificial sweetener that does retain its flavor when cooked.

When you hear claims that are supported only by a fringe minority that's in opposition to the scientific consensus, you have good reason to be skeptical right off the bat, but it doesn't mean it's not worth looking into. Aspartame has been looked into ad nauseam even after its approval, and found safe at every try; so at some point you have to depart from rationality to continue supporting the claims made against it. Enjoy your diet Dr. Pepper, it's not going to hurt you; if it was, I'd have been dead decades ago.

References & Further Reading

Aaronovitch, D. *Voodoo History: The Role of the Conspiracy Theory in Modern History.* New York: Riverhead, 2010.

Butchko, H., Stargel, W., Comer, C., Mayhew, D., Benninger, C., Blackburn, G., de Sonneville, L., Geha, R., Hertelendy, Z., Koestner, A., Leon, A., Liepa, G., McMartin, K., Mendenhall, C., Munro, I., Novotny, E., Renwick, A., Schiffman, S., Schomer, D. "Aspartame: review of safety." *Regulatory Toxicology and Pharmacology.* 1 Apr. 2002, Volume 35, Number 2 Pt 2: S1-93.

Magnuson, B.A., Burdock, G.A., Doull, J., Kroes, R.M., Marsh, G.M., Pariza, M.W., Spencer, P.S., Waddell, W.J., Walk.er, R., Williams, G.M. "Aspartame: a safety evaluation based on current use levels, regulations, and toxicological and epidemiological studies." *Critical Reviews in Toxicology.* 1 Jan. 2007, Volume 37, Number 8: 629-727.

Rulis, A.M., Levitt, J.A. "FDA's food ingredient approval process: Safety assurance based on scientific assessment." *Regulatory Toxicology and Pharmacology.* 1 Feb. 2009, Volume 53, Number 1: 20-31.

Stegink, L.D. "The aspartame story: a model for the clinical testing of a food additive." *The American Journal of Clinical Nutrition.* 1 Jul. 1987, Volume 46, Number 1: 204-215.

8. Scalar Weapons: Tesla's Doomsday Machine?

No incoming missiles or airplanes warn of the hellish destruction that is almost upon us. No bombs fall, no hidden explosives tick away ominously. People go about their daily lives on a beautiful day. When suddenly, every molecule within 50 miles leaps to a temperature hotter than the sun, vaporizing every living creature, shattering every rock into plasma, obliterating every construction. The atmosphere flashes to dissociated gases and expands to many times its size, kicking off a terrible shockwave that rips three times around the planet. The awful event lasts less than a second, and not even dirt remains behind. Ash and slag rain for several hours. This is a scalar weapon, a power unlike anything in conventional physics, said to be invented by the obsessive genius Nikola Tesla.

Some point to the 1908 Tunguska event as Tesla's own proof of concept test. A series of unexplained booms off the eastern coast of the United States in 1977, and one over the Netherlands in 1976, have been attributed to Soviet testing of a scalar weapon. Some believe that the Japanese *yakuza* crime organization regularly uses scalar technology to create typhoons and other weather events throughout Asia. So what is this "scalar" stuff, how does it work, is it real, is it something we need to worry about, and what has Nikola Tesla got to do with it?

Let's start off with some basic definitions. A *scalar field* is a concept in mathematics and physics, in which a single value is assigned to every point in space. An example of this would be to describe the temperature in space, where there would be a single finite number assigned to every point. The science behind these is called scalar field theory. Compare this to a *vector field*, in which every point in space has a vector, consisting of a direction and a strength. Gravity in space can be defined by a vector field, as can the magnetic field surrounding a magnet. Those are legitimate science. Many of the same terms are used by the proponents of a pseudoscientific version also called *scalar field theory*, and from now on, whenever I refer to scalar field theory we're talking about the made-up version. This new type of scalar field theory takes it a step beyond legitimate science, based on the assumption that the scalar field has four or more dimensions, in which something they call *scalar energy* is also present at each point in space. *Scalar waves* are the hypothetical electromagnetic waves propagating along this field; although, unlike conventional waves that propagate outward like ripples in a pond, scalar waves propagate through space longitudinally, like ocean waves breaking on a long straight beach. These scalar waves, also called *Tesla waves* or *Maxwellian waves*, are said to be the mechanism of zero-point energy. It should be stressed that this definition of scalar field theory is not supported by experiment or by any actual physics.

This is fortuitous for the proponents of scalar field theory: Once you leave the realm of real science, you can pretty much make up whatever you want, and it's no more or less legitimate. There are endless web pages dedicated to this topic, full of meaningless technobabble that cram in so many scientific sounding terms that a layperson has no hope of discriminating science from pseudoscience.

The basic concept of a *scalar weapon* opposes two powerful scalar waves against each other, creating a hypothetical standing wave resulting in what they call a *scalar bubble* between them. By controlling the strength and location of this scalar bubble in space, it is supposed to superheat the target area (though it's not clear why), even to the point of vaporizing the atmosphere itself. An enemy could theoretically enclose New York City within a giant scalar bubble, flashing it into oblivion.

For the sake of Nikola Tesla's legitimate legacy, it is important to separate his name from the modern pseudoscience of scalar field theory. It is true that Tesla did envision and describe superweapons capable of frying entire invading armies, but his concept was analogous to what we now call a directed energy weapon, basically a powerful particle beam. He did not stray into nonsense like scalar bubbles. Tesla did also claim to have completed a partial unified field theory that unified gravity with electromagnetism, which is something that scalar field theory also claims. Because of these similarities, Tesla's name is often wrongly associated with scalar weapons and scalar field theory. In many cases, his name is outright hijacked by scalar weapon proponents and used to give their ideas the appearance of credibility.

I should probably take a moment to defend my use of terms like "nonsense" to describe scalar bubbles and such things. Since these are theoretical and not evidence based, there's no way their existence can be disproven. Every so-called demonstration of scalar wave effects can also be fully attributed to the known effects of electrodynamics. In short, it's equally valid to say that the scalar field has a fourth dimension consisting of Mickey Mouses at every point in space, who all pull on ropes, thus creating all the fundamental forces in physics. This is equally non-falsifiable, and would produce the same experimental results as those we hear from the scalar field theory crowd.

The principal public proponent of pseudoscientific scalar field theory is Thomas Bearden, a retired Lt. Colonel in the U.S. Army. Most web articles about scalar weapons cite Bearden's writing as their principle source. He has often written under the guise of a Ph.D. purchased from an unaccredited "life experience" diploma mill. Most of his many books, papers, and web sites are about perpetual motion machines, free energy, magnetic motors, and other "over-unity" violations of the laws of thermodynamics. Among his claims are that scalar weapons and other such technologies are responsible for Chernobyl, the destruction of the space shuttle *Challenger,* the downing of TWA Flight 800, the 2004 Indian Ocean tsunami, and Hurricane Katrina. You might be surprised to hear it, but I actually like Bearden. His science is largely fantasy based, but he seems a good genuine guy who hopes that scalar technologies will benefit humanity. On one page of his web site he shows a pile of his books about to be shipped out, and he comments "There's the information, on its way to going out, and perhaps to the very grad student who eventually turns the academic energy world upside down and makes it happen." That's a hope a lot of us share, but the cold hard reality is that scientific progress is almost always the result of long, hard, tedious work, and rarely a fortuitous sudden rewriting of the rules from the fringe. Bearden's profound and uncritical belief in nearly every conspiracy theory imaginable is fairly typical among many proponents of scalar weapons, and it clearly clouds their judgment.

Bearden notes that a letter he received from the National Science Foundation, in response to an email describing his perpetual motion machines, states "There is a uniform support for your 'out of the box' thinking about conventional models and mathematical approaches." What's often omitted by Bearden's supporters is that this letter diplomatically concludes by emphasizing the need to "(1) demonstrate the strength of evidence that perpetual motion machines have worked as advertised, and (2) address how something works that appears to violate our present understanding of engineering and physics." He's wrong and has a lot of pretty crazy theories, but so do a lot of other good people. It's great to be well intentioned, but it's also equally important we better inform ourselves before propagating misinformation. And this is the reason to quarrel with Bearden. He is very well informed, but about a fantastical, nonscientific universe.

Just Google for "scalar weapon" and you'll find more than enough reading material to keep you occupied for days. One interesting trend to watch for is the frequent use of the terms "old" and "new": The "old" understanding of physics and electromagnetism, and the "new" understanding. Make no mistake; "old" and "new" physics really mean "real" and "made up" physics. You'll see that virtually every authoritative link or reference is to one of Tom Bearden's books or web pages. You'll see all the familiar warning signs of the classic conspiracy mindset: Huge lists linking nearly every aerospace or weather-related disaster to scalar weapons, and the uncritically presumed existence of worldwide networks of secret weaponry, men in black, and confessions of anonymous insiders claiming that such things are real.

The first time most people hear about scalar weapons is usually through a YouTube video or chain email from some doomsayer. Whenever you hear such a wild, far-out story, you should always approach it with skepticism, and not just accept it at face value because the chain email was forwarded by a trusted friend. The proponents of scalar weapon conspiracy theories are not backed by any valid science. The idea makes for some fine science fiction, but at a minimum, spend five minutes on Wikipedia before accepting and repeating such wild stories as science fact.

References & Further Reading

Bearden, T. *Analysis of Scalar/Electromagnetic Technology*. Chula Vista: Tesla Book Co., 1990.

Durrant, A.V. *Vectors in Physics and Engineering*. London: Chapman & Hall, 1996. 127-140.

Melia, Fulvio. *Electrodynamics*. Chicago: The University of Chicago Press, 2001.

Nichelson, Oliver. "Tesla's Wireless Power Transmitter and the Tunguska Explosion of 1908." *Prometheus Alternative Sciences and Technology*. Prometheus Group, 8 Jul. 2001. Web. 30 Sep. 2008. <http://prometheus.al.ru/english/phisik/onichelson/tunguska.htm>

Tesla, N. *My Inventions: The Autobiography of Nikola Tesla*. Williston, VT: Hart Bros., 1982.

9. HAARP MYTHS

Hold on tight, because the U.S. Government is using HAARP, the High Frequency Active Auroral Research Program in Alaska. Some charge that this secret government project can modify the weather, creating hurricanes and typhoons; that it can create earthquakes and superheat the atmosphere; or even that it can destroy aircraft anywhere in the world; and control the minds of its victims. What do we say when we hear stories that sound far fetched or implausible? *Be skeptical.*

Let's talk about the claims made about HAARP, but first let's talk about what it actually is and what exactly it's really capable of. First of all, there's nothing remotely secret or even classified about HAARP. No security clearance is needed to visit and tour the site, and HAARP usually holds an open house every summer during which anyone can see everything there. During the rest of the year, research is conducted. The universities that have participated in HAARP research include University of Alaska, Stanford, Penn State, Boston College, Dartmouth, Cornell, University of Maryland, University of Massachusetts, MIT, Polytechnic University, UCLA, Clemson and the University of Tulsa. There are several other similar research stations around the world, namely the Sura facility in Russia, EISCAT in Norway, the Arecibo observatory in Puerto Rico, and the HIPAS observatory near Fairbanks, operated by UCLA. If you look at HAARP on Google Earth, you can see there's not much there, and the current view shows only four cars in the small parking lot.

HAARP consists of an observatory and an adjacent 28-acre field with 180 HF (high frequency) antennas, each 72 feet tall, with a maximum transmission power of 3600 kilowatts, about 75 times the power of a commercial radio station, but only a tiny fraction of the strength of the natural solar radiation striking the same part of the ionosphere at which HAARP is aimed. Although the observatory operates continuously, the HF antenna array is activated only rarely for specific experiments, which average about once a month.

Sadly for the conspiracy theorists, HAARP has no potential to affect weather. The frequency of energy that HAARP transmits cannot be absorbed by the troposphere or the stratosphere, only by the ionosphere, many kilometers higher than the highest atmospheric weather systems.

The ionosphere is created and replenished daily by solar radiation. At night, the level of ionization drops quickly to very low levels at lower altitudes of 50 to 100 miles, but at higher altitudes over 200 miles it takes most of the night for the ionization to disperse. During the night, when the natural ionosphere is minimal, HAARP is capable of creating a weak artificial aurora that can actually be observed by sensitive cameras at the observatory, though they are far too faint for the naked eye. During the daytime, solar radiation ionizes the ionosphere so powerfully that HAARP's weak artificial effects are the proverbial drop in the bucket, and are erased almost immediately when the transmitter is turned off.

You might ask "What's the point of HAARP?" If it's not to wreak global destruction, what good is it? Communication and navigation signals are sent through the atmosphere for a broad range of civilian and military purposes. Guided missiles rely on digital transmissions which can be affected or jammed by a whole variety of natural and artificial causes. Global Positioning System and encrypted communications all need to be able to make it to their recipients in wartime, regardless of the atmospheric and electromagnetic conditions. The study of these effects is the primary reason that DARPA, the U.S. Air Force, and the U.S. Navy contribute to HAARP's funding. In addition, by bouncing signals off the ionosphere at an altitude of 100km, HAARP has been able to create Extremely Low Frequency, or ELF, waves as low as 1 Hertz, which can potentially be used for worldwide communication including reaching submarines, though at an almost uselessly slow data rate. But before you conclude that these ELF waves might be used for creating earthquakes, note that the maximum ELF signal amplitude produced by HAARP has been measured at less than one ten-millionth of the Earth's natural background field.

So if HAARP is so anticlimactically mundane, why all the conspiracy theories? HAARP is operated by MarshCreek, LLC, an Alaska Native Corporation under contract to the Office of Naval Research. Anytime the ONR or DARPA or the military have their hand in something, paranoid types tend to come out of the woodwork and blame anything they can imagine on it. So regardless of whether HAARP is in the atmospheric research business or the rubber duckie business, they were pretty much doomed to conspiracy charges from the beginning.

But there is also a secondary reason that HAARP has been suspected of deeper, darker purposes, and it goes back to its early construction. The winning contractor to build HAARP was ARCO Power Technologies, or APTI. ARCO has historically been one of Alaska's largest employers and they initially set up APTI as a subsidiary to construct power plants using Alaska's vast natural gas reserves. One scientist employed at APTI was Dr. Bernard Eastlund, a physicist of some note. Among Dr. Eastlund's accomplishments was the co-invention of the fusion torch, and the original owner of a 1985 U.S. patent on a "Method and apparatus for altering a region in the earth's atmosphere, ionosphere, and/or magnetosphere." Dr. Eastlund's method required a location near the poles, where the lines of the Earth's magnetic field are more or less perpendicular to the surface, like Alaska, and presumed a natural gas power source. A few years later, the HAARP program began. A coincidence? No way, say the conspiracy theorists.

It seems logical to me that if I were ARCO and wanted to get in on a lucrative government construction contract and sell them my Alaskan natural gas, I might well set up a subsidiary with one of the world's leading experts in the field. To me this looks like a smart business move by ARCO and by the government; how it would suggest an evil conspiracy to destroy the world, I'm just not seeing that.

Dr. Eastlund's patent, which has since become popularly known (though inaccurately) as the "HAARP patent", is widely reproduced online, often with much commentary from authors making their own interpretations of how it might be used. Specifically, the patent involves using natural gas to generate electricity to create electromagnetic radiation to excite a tiny section of the ionosphere to about 2 electron volts, thus moving it upward along the lines of the magnetic field. The conspiracy theorists, once again, completely ignore the fact that this can only happen in the ionosphere, and they interpret it as a weather control system or earthquake generating system. Such extrapolations are without any plausible foundation.

A further disconnect in this conspiracy claim is that Dr. Eastlund's patent was for a speculative and unproven device approximately one *million* times as powerful as HAARP. The patent does not mention HAARP, and none of its drawings remotely resemble anything built at HAARP. For perspective, HAARP's antenna array measures about 1000 feet on a side. A device such as that imagined by Dr. Eastlund would have been 14 *miles* on a side, with one million antenna elements, compared to HAARP's 180. Furthermore, Dr. Eastlund left APTI to found his own company before the HAARP program began, and was never associated with the program.

One of the most vocal critics of HAARP is Nick Begich, son of the late Alaskan congressman of the same name. He writes as Dr. Nick Begich, but his Ph.D. is in traditional medicine and was purchased via mail from the unaccredited Open International University in India, and included no coursework or curriculum. Begich is a proponent of a number of new age energy healing techniques of his own invention. In 1995 he self-published *Angels Don't Play This HAARP*. This book kick-started many of the popular rumors about HAARP, including that mass mind control is one of its goals.

A conspiracy theorist named Benjamin Fulford has made some YouTube videos charging that HAARP is responsible for most of the severe earthquakes around the world, and that the United States threatens nations like Japan with earthquakes if they don't "do what we want". He believes that HAARP accomplishes this by heating up water in the atmosphere the same way that a microwave oven does, though he is not clear on how warming a tiny patch of upper atmosphere in Alaska would cause an earthquake in Asia with pinpoint precision. There is no known correlation between temperature and earthquakes. Fulford's microwave theory is also wide of the mark. HAARP's maximum frequency is 10 MHz, and the dielectric heating effect of a microwave oven requires 2.5 GHz, or 250 times higher than HAARP. Dielectric heating also requires reversing the polarity of the field more than a million times a second, one thousand times HAARP's fastest frequency. A note to conspiracy theorists: At least *pretend* to know what you're talking about.

Fulford bolsters his claim with some beautiful video of dramatically illuminated clouds, which he calls "earthquake lights" and believes constitutes evidence that HAARP caused the 2008 Sichuan earthquake in China. In fact these are simply clouds illuminated by the sun after it has dipped below the horizon, and are quite common.

One of the more colorful HAARP conspiracy theorists is a woman on YouTube who goes by the name "dbootsthediva", lately more commonly known as "the crazy sprinkler lady". Her YouTube page contains about 50 videos she has made of her house and yard, with her making commentary about how HAARP is responsible for virtually every little thing she sees — everything from a rainbow in her sprinklers, to a moiré pattern on the clapboard siding of her house, to moving the ground under her feet and causing the picture to shake. She manages to see HAARP's alleged affects wherever and whenever she looks, despite the fact that HAARP is rarely actually transmitting.

But I would go blue in the face long before I could describe even a fraction of all the bizarre fears about HAARP trumpeted on the Internet. By now I've learned there's no hope of changing the minds of some people who have latched onto the idea that global domination is as easy as the erection of what amounts to little more than 180 cell phone towers — if destroying another country was this trivial, you'd think America's enemies would have done it to us long ago. DARPA has its hand in many research projects, like robotics and the Grand Challenge autonomous vehicle races, all of which have civilian as well as military applications and all of which represent good science. When you hear the assumption that just because DARPA funds something it must automatically be an evil superweapon, you have good reason to be skeptical.

References & Further Reading

Eastlund, Bernard J. "Method and apparatus for altering a region in the earth's atmosphere, ionosphere, and/or magnetosphere." *United States Patent Database.* Patent US, 11 Aug. 1987. Web. 9 Jan. 2010. <http://patft.uspto.gov/netacgi/nph-Parser?Sect1=PTO1&Sect2=HITOFF&d=PALL&p=1&u=/netahtml/PTO/srchnum.htm&r=1&f=G&l=50&s1=4,686,605.PN.&OS=PN/4,686,605&RS=PN/4,686,605>

HAARP. "HAARP Fact Sheet." *The High Frequency Active Auroral Research Program.* HAARP Alaska, 15 Jun. 2007. Web. 9 Jan. 2010. <http://www.haarp.alaska.edu/haarp/factSheet.html>

Inan, U. S., Bell, T. F. "Polar Aeronomoy and Radio Science (PARS) ULF/ELF/VLF Project." *Stanford University Star Laboratory.* Star.Stanford, 1 Jul. 2001. Web. 9 Jan. 2010. <http://www-star.stanford.edu/~vlf/pars/pars.htm>

Pike, John. "Extremely Low Frequency Communications Program." *Federation of American Scientists.* FAS, 26 Feb. 2004. Web. 12 Jan. 2010. <http://www.fas.org/nuke/guide/usa/c3i/elf.htm>

Ratcliffe, John Ashworth. *An introduction to the ionosphere and magnetosphere.* London: Cambridge University Press, 1972.

10. Ten Most Wanted: Celebrities Who Promote Harmful Pseudoscience

Today I have a list for you: The ten celebrities who most abuse their fame to promote dangerous or otherwise harmful misinformation. You may be disappointed that this is not simply a list of Hollywood Scientologists. On the contrary, I think Tom Cruise deserves a medal. He's done more to discredit Scientology than anyone else. If anyone didn't already think Scientologists were nuts, Tom Cruise has sealed the deal. You also won't find anyone who's simply a harmless wacko. I endeavored to include only celebrities who are actively doing harm to the public by spreading misinformation that does damage.

#10 - Montel Williams

He's all the way down at the bottom of the list because his daytime talk show is no longer on the air and he doesn't have much influence anymore, but when he did, he was best known for promoting psychics as the best way to solve almost any crisis. You can quarrel with psychic predators like Sylvia Browne, but her career was created by Montel Williams. Montel's worst offense was to use psychics to provide made-up information to the parents of missing children, which he did on many occasions, not just the one or two high profile cases that made headlines. Without exception, this information has always been either uselessly general or flat-out wrong. All the while, Montel Williams unapologetically promoted psychic powers to his millions of viewers.

#9 - Chuck Norris

He deserves to be on the list anyway for making nothing but stupid movies, but Chuck Norris' main offense is his frequent public appeals to teach a Biblical "alternative" to science in public schools. In a series of public service announcements, Chuck and his wife advocate the mission of the National Council on Bible Curriculum in Public Schools, a nonprofit organization with its own 300 page textbook advocating Young Earth fundamentalism, *The Bible in History and Literature*. Although Chuck and the Council state that it's legal and has never been legally challenged, this is patently untrue, its having failed every Constitutional challenge brought forth against it. Chuck, become a Sunday School teacher in the church of your choice. You should not use your celebrity status to wage war against religious freedom, or to further erode the quality of science education in the United States.

#8 - Joe Rogan

Comedian Joe Rogan does what he can to promote virtually any conspiracy theory that he stumbles onto, apparently accepting them all uncritically with a wholesale embrace. He believes the Apollo astronauts did not land on the moon. He believes the U.S. government was behind the 9/11 terrorist attacks. He believes the Oliver Stone version of the Kennedy assassination. He believes aliens crashed at Roswell in 1947 and the government is covering it up. He thinks Men in Black from Project Blue Book stole his friend's camera, even though Project Blue Book ended over 38 years ago. The worst part is that he promotes these ideas to the public at every interview opportunity, but gives himself the intellectual "Get out of jail free" card of not needing any evidence by hiding behind the childish debate technique of saying "Hey, I'm just the guy asking questions." Joe, if you're going to put so much effort into promoting conspiracy theories and eroding what little rationality the public has left, at least have the courage to come forward with a cogent argument and well-sourced evidence, instead of the lameness of "I'm just the guy asking questions." Take the responsibility.

#7 - Ben Stein

There's nothing wrong with being a religious person, but actor Ben Stein takes it many steps further, employing fallacious logic to claim that everything bad in the world is caused by non-Christian ideas. His favorite is that the study of science caused the Holocaust. He's now infamous for his quote "the last time any of my relatives saw scientists telling them what to do they were telling them to go to the showers to get gassed … that's where science leads you." Ben's open hostility toward scientific literacy is aptly described by former *Scientific American* Editor-in-Chief John Rennie, who wrote: "Ben Stein wants you to stop thinking of evolution as an actual science supported by verifiable facts and logical arguments and to start thinking of it as a dogmatic, atheistic ideology akin to Marxism." Science is, quite properly, independent of politics and religion. A celebrity who argues that science should be subservient to either, especially one who exploits the Holocaust to do so, is an intellectual felon.

#6 - Pamela Anderson

Although we here at Skeptoid endorse their annual "Running of the Nudes" in Pamplona, Spain, we don't like anything else about PETA, People for the Ethical Treatment of Animals. Pamela Anderson lends her celebrity to them and serves as one of their primary spokespeople, as do many other celebrities. PETA has been widely criticized for its support of self-described domestic terrorist groups Earth Liberation Front and Animal Liberation Front. Groups like PETA do far more harm than good to the animal rights movement by exploiting the Holocaust for its advertising or for complaining only about the death of a donkey in a Jerusalem bomb attack that killed dozens of people. And Pamela, you might want to think twice before donating money to PETA. The Better Business Bureau's Wise Giving Alliance has noted that PETA fails to meet several Charity Accountability standards, and a Senate committee has questioned its tax exempt status for funding organizations later designated as terrorist.

#5 - Larry King

Larry King's job as a professional interviewer is to bring on a huge number of people from all backgrounds and let them speak their minds, and this is a good thing. We hear from people doing good, people doing bad, people we agree with, and people we disagree with. But Larry's show is supposed to be better than all the other interview shows. Only Larry gets to talk to heads of state, U.S. Presidents, the top movers and shakers. He hits them hard, asks them the tough questions, puts them on the spot. Unless — and that's a very big unless — they are on the show to promote some pseudoscience or paranormal claim. Of these guests, Larry asks no tough questions. He gives them an unchallenged platform to promote their harmful claim. He gives their web addresses and shows their books and DVDs. He acts as their top salesman for the hour. Larry King gives every indication that CNN fully endorses celebrity psychics, conspiracy theorists, ghost hunters, UFO advocates, and promoters of non-scientific alternatives to healthcare.

#4 - Bill Maher

While we love Bill Maher's movie *Religulous* and appreciate that his is one of the very few public voices opposing the 9/11 conspiracy myths, we can't deny that he has a darker side. Bill Maher is a board member of PETA — one of the people actually approving their payments to people like convicted arsonist Rod Coronado — but his ongoing act that's most harmful to the world is his outspoken denial of evidence-based medicine.

Yes, Bill is correct that a good diet and exercise are good for you, but he seems to think that doctors deny this. Not any doctor I've ever spoken to. Bill made it clear on a four-minute speech on his show that he believes government and Big Pharma conspire to keep everyone sick by prescribing drugs. If even a single person takes Bill's claims to heart and avoids needed medical treatment as a result, Bill Maher is guilty of a terrible moral crime. Considering the huge size of his audience, this seems all too likely.

#3 - Prince Charles

What's even worse than a comedian denying modern medicine is when the future King of England does the same thing. This is the kind of medieval superstition we expect from witch doctors like South Africa's former health minister Manto Tshabalala-Msimang, not from the royal family of one of the world's most advanced nations (well, it would be, except that royal families are kind of a medieval thing too). Through The Prince's Foundation for Integrated Health (which has since shut down amid scandal), Prince Charles attempts to legitimize and promote the use of untested, unapproved, and implausible alternative therapies of all sorts instead of using modern evidence-based medicine. He has a "collaborative agreement" with Bravewell, the United States' largest fundraising organization dedicated to the promotion of non-scientific alternatives to healthcare. As perhaps the most influential man in the United Kingdom, Prince Charles displays gross irresponsibility that directly results in untreated disease and death.

#2 - Jenny McCarthy

The most outspoken anti-vaccine advocate is, by definition, the person responsible for the most disease and suffering in our future generation. Jenny McCarthy's activism has been directly blamed for the current rise in measles. She also blames vaccines for autism, against all the well-established evidence that shows autism is genetic, and she spreads this misinformation tirelessly. She believes autism can be treated with a special diet, and that her own son has been "healed" of his autism through her efforts. Since one of the things we do know about autism is that it's incurable, it seems likely that her son probably never even had autism in the first place. So Jenny now promotes the claim that her son is an "Indigo child" — a child with a blue aura who represents the next stage in human evolution. If you take your family's medical advice from Jenny McCarthy, this is the kind of foolishness you're in for. Instead, get your medical advice from someone with a plausible likelihood of knowing something about it, like say, oh, a doctor, and *not* a doctor who belongs to the anti-vaccine Autism Research Institute or its Defeat Autism Now! project. Go to StopJenny.com for more information.

#1 - Oprah Winfrey

The only person who can sit at the top of this pyramid is the one widely considered the most influential woman in the world and who promotes every pseudoscience: Oprah Winfrey. To her estimated total audience of 100 million, many of whom uncritically accept every word the world's wealthiest celebrity says, she promotes the paranormal, psychic powers, new age spiritualism, conspiracy theories, quack celebrity diets, past life regression, angels, ghosts, alternative therapies like acupuncture and homeopathy, anti-vaccination, detoxification, vitamin megadosing, and virtually everything that will distract a human being from making useful progress and informed decisions in life. Although much of what she promotes is not directly harmful, she offers no distinction between the two, leaving the gullible public increasingly and incrementally injured with virtually every episode.

When you have a giant audience, you have a giant responsibility. Maybe you don't want such a responsibility, in which case, fine, keep your mouth shut; or limit your performance to jokes or acting or whatever it is you do.

References & Further Reading

Bartholomaus, Derek. "Jenny McCarthy Body Count." *Jenny McCarthy Body Count.* Derek Bartholomaus, 12 Dec. 2009. Web. 27 Dec. 2009. <http://www.jennymccarthybodycount.com>

Bidlack, Hal. "An Open Letter to Lt. Commander Montel Williams." *Stop Sylvia Brown.* Stop Sylvia, 6 Feb. 2007. Web. 28 Dec. 2009. <http://stopsylvia.com/articles/openlettertomontel.shtml>

Morrison, Adrian R. "Personal Reflections on the "Animal-Rights" Phenomenon." *The Physiologist.* 1 Feb. 2001, Volume 44, Number 1: 1.

Noveck, Jocelyn. "Somers' New Target: Conventional Cancer Treatment." *ABC News Health.* ABC News, 19 Oct. 2009. Web. 28 Dec. 2009. <http://abcnews.go.com/Health/wirestory?id=8866956&page=1>

Rennie, J., Mirsky, S. "Six Things in Expelled That Ben Stein Doesn't Want You to Know..." *Scientific American.* Scientific American, 16 Apr. 2008. Web. 28 Dec. 2009. <http://www.scientificamerican.com/article.cfm?id=six-things-ben-stein-doesnt-want-you-to-know>

Singh, S., Ernst, E. *Trick or Treatment, The undeniable facts about alternative medicine.* New York: Bantam Press, 2008.

11. Betty and Barney Hill: The Original UFO Abduction

It was shortly before midnight on September 19, 1961 when Betty and Barney Hill had the experience that was to shape all of modern alien folklore. They were driving from Canada to Portsmouth, New Hampshire. Near the resort of Indian Head, New Hampshire, they stopped their car in the middle of Route 3 to observe a strange light moving through in the sky. The next thing they knew, they were about 35 miles further along on their trip, and several hours had elapsed.

Betty telephoned their close friend, Major Paul Henderson at nearby Pease Air Force Base, to report a UFO sighting. Major Henderson found that this was corroborated by two separate UFO reports from radar data from two different Air Force installations nearby. All three reports are officially recorded in Project Blue Book. Then Betty began having nightmares two weeks later. In her nightmares, she described being taken aboard an alien spacecraft and having medical experiments performed. As a result of these nightmares, Betty and Barney decided to undergo hypnosis. In separate sessions, they described nearly identical experiences of being taken on board the alien spacecraft by what we now call gray aliens: Short beings with huge black eyes and smooth gray skin. Both of the Hills had a whole spectrum of tests done. Betty was shown a star map that she was able to memorize and reproduce later, and which has been identified as showing Zeta Reticuli as the aliens' home planet. After the experiments they were taken back to their car in a dazed condition, and sent along their way.

Innumerable books and movies were made about the Betty & Barney Hill abduction. It was the introduction of the gray alien into popular culture. It was also the beginning of the entire "alien abduction" phenomenon. The physical evidence of the star map and the radar reports are said to have both withstood all scrutiny. In fact you almost never hear a critical treatment of their story.

Much of the Hill story is said to be based on these separate hypnosis sessions. In fact, that turns out not to be the case at all. It's important to note that it was more than two years after the incident that the Hills underwent hypnosis. During those two years, Betty was writing and rewriting her accounts of her dreams. All of the significant details you may have heard about the Hills' medical experiments came from Betty's two years of writings: A long needle inserted into her navel; the star map; the aliens' fascination with Barney's dentures; the examination of both Betty and Barney's genitals; and the overall chronology of the episode, including being met on the ground by the aliens, a leader coming forward and escorting them to exam rooms, the aliens' general demeanor and individual personalities, and the way they spoke to Betty in English but to Barney via telepathy. Betty wrote all of this based only on what she claims were her dreams, and probably told the story to Barney over and over again until his ears fell off over a period of two years, before they ever had any hypnosis.

During those two years, Barney's own recollection was somewhat less dramatic. When they first saw the light in the sky, Betty said she thought it was a spacecraft, but Barney always said he thought it was an airplane.

Betty's written description of the characters in her nightmare depicted short guys with black hair and "Jimmy Durante" noses. It was only in Barney Hill's hypnosis sessions that we got the first description of skinny figures with gray skin, large bald heads, and huge black eyes. After Betty Hill heard these sessions, suddenly her own hypnosis accounts began to describe the same type of character, and from that moment on, she never again mentioned her original Jimmy Durante guys. Many modern accounts wrongly state that her original nightmares also described grays.

Although the popular version of events is that Barney Hill's hypnosis description is the first appearance of a so-called gray alien in modern culture, that first appearance actually came twelve days earlier, on national television, in an episode of *The Outer Limits* called *The Bellero Shield*. The alien in that episode shared most of the significant physical characteristics with the alien in Barney's story: Bald head, gray skin, big wraparound eyes. The Hills stated they did not watch it and didn't know about it.

Remember: Before examining the specific claims made in a fantastic story, you should check the source of the story for credibility. Barney Hill died only a few years after the alleged incident, but Betty Hill stuck around long enough for her credibility to be pretty thoroughly demonstrated. *Skeptical Inquirer* columnist Robert Shaeffer wrote:

> *I was present at the National UFO Conference in New York City in 1980, at which Betty presented some of the UFO photos she had taken. She showed what must have been well over two hundred slides, mostly of blips, blurs, and blobs against a dark background. These were supposed to be UFOs coming in close, chasing her car, landing, etc... After her talk had exceeded about twice its allotted time, Betty was literally jeered off the stage by what had been at first a very sympathetic audience. This incident, witnessed by many of UFOlogy's leaders and top activists, removed any lingering doubts about Betty's credibility — she had none. In the oft-repeated words of one UFOlogist who accompanied Betty on a UFO vigil in 1977, she was "unable to distinguish between a landed UFO and a streetlight." In 1995, Betty Hill wrote a self-published book, A Common Sense Approach to UFOs. It is filled with obviously delusional stories, such as seeing entire squadrons of UFOs in flight and a truck levitating above the freeway.*

She also once wrote in a 1966 letter "Barney and I go out frequently at night for one reason or another. Since last October, we have seen our 'friends' on the average of eight or nine times out of every ten trips." But is it possible that Betty's obsession with UFOs could have been caused by her trauma from a genuine abduction? Yes, it's possible that it could have pushed her further in that direction, but Betty had commonly spoken of UFOs even before 1961, including one story she often told of her sister's own close encounter in 1957.

So here's what we have so far: A woman who clearly had an obsession with UFOs saw a light in the sky that her husband described as an airplane. She then spent two years writing an elaborate story and no doubt telling and retelling it to her husband. Later, under hypnosis, Barney was asked about the events described in Betty's story, and surprise surprise, he retold the story she'd already told him a hundred times, with an added dash from *The Outer Limits* episode of twelve days before. So far, we have a tale that's hard to consider reliable.

But then there are those three items said to be physical evidence of the Hill abduction: first, the star map hand drawn by Betty by memory from one shown to her aboard the spacecraft; second, the purple dress she was wearing on that night, kept for forty years in her closet, torn and covered with mysterious dust; and third, reports in the Air Force's official Project Blue Book stating that radar confirmed the presence of a UFO on that night at two separate Air Force facilities in the area, both within hours of the Hills' claimed abduction. Let's look at those first.

The first report was from Pease Air Force Base, about 82 miles southeast of Indian Head, at 2:14am. The Hills got home in Portsmouth at 5:00 in the morning on September 20. Their story states that they came to after their medical experiments about 35 miles south of Indian Head, near the town of Ashland. From Ashland to Portsmouth is about an hour and 45 minute drive, so they came to in their car around 3:15. This chronology puts Pease AFB's UFO radar evidence squarely in the middle of the Hills' three hours of medical experiments aboard the spaceship, which they say was sitting on the ground the whole time. If the Hills' story is true, the Pease AFB report *must* be an unrelated event.

The second report is from North Concord Air Force Station, a small hilltop radar station (closed in 1963) that was about 40 miles north of Indian Head, at 5:22pm on September 19. This is about seven hours before the Hills observed their light in the sky. It clearly does not corroborate the Hills' sighting. The reports in Project Blue Book note each target's extremely high altitude and low speed, and conclude that each was probably a weather balloon.

Next we have Betty's purple dress, the zipper of which she found to be torn. She then hung it in the closet. Two years later, after the hypnosis, she got it out and said there was strange pink dust on it. She hung it up again, this time for forty years, when a group of crop circle investigators examined it. They concluded the dress had an "anomalous biological substance" on it. While a good stretch of the imagination might consider this to be consistent with the abduction story, it's also consistent with perfectly natural explanations, namely, 40 years of dust mites, moths, and mold. I don't find the Great Purple Dress Caper to be good evidence of anything.

So the only thing we're left with is Betty's star map. In her original written stories, she described the aliens' star map as three dimensional. Under hypnosis, she redrew it on paper, in two dimensions. It's seven or eight random dots connected by lines, and it's quite rough and by no means precise. Several years later, a schoolteacher named Marjorie Fish read a book about the Hills. She then took beads and strings and converted her living room into a three dimensional version of the galaxy based on the 1969 Gliese Star Catalog. She then spent several *years* viewing her galaxy from different angles, trying to find a match for Betty's map, and eventually concluded that Zeta Reticuli was the alien home world. Other UFOlogists have proposed innumerable different interpretations. Carl Sagan and other astronomers have said that it is not even a good match for Zeta Reticuli, and that Betty's drawing is far too random and imprecise to make any kind of useful interpretation. With its third dimension removed, Betty's map cannot contain any useful positional information. Even if she had somehow drawn a perfect 3D map that did exactly align with known star positions, it still wouldn't be evidence of anything other than that such reference material is widely available, in sources like the Gliese Star Catalog. We would not conclude that an alien abduction is the only reasonable way that Betty could have learned seven or eight star positions during those two years.

And so, there we have it. The Betty & Barney Hill abduction story has every indication of being merely an inventive tale from the mind of a lifelong UFO fanatic. Despite the best efforts of authors to bolster it with mischaracterized or exaggerated evidence, it is unsupported by any useful evidence, and is perfectly consistent with the purely natural explanation.

REFERENCES & FURTHER READING

Fuller, John G. *The Interrupted Journey: Two Lost Hours Aboard a Flying Saucer.* New York: Dell, 1966.

Klass, Philip J. *UFO-Abductions: A Dangerous Game.* Buffalo: Prometheus Books, 1988. 7-15.

National Archives and Records Administration. "16-30 September 1961 Sightings." *Project Blue Book.* Project Blue Book, 1 Sep. 2005. Web. 19 Jan. 2010. <http://www.bluebookarchive.org/page.aspx?pagecode=NARA-PBB1-259&tab=1>

Pflock, Karl and Brookesmith, Peter, editors. *Encounters at Indian Head: The Betty and Barney Hill UFO Abduction Revisited.* San Antonio: Anomalist Books, 2007.

Sheaffer, Robert. *The UFO Verdict: Examining the Evidence.* Buffalo: Prometheus Books, 1986. 34-44.

12. The Incorruptibles

Lightning flashes as we scrape the final shovelfuls of dirt away from the top of the coffin, pry open the lid, and in the lantern light we see a perfectly preserved human body! It's as if she died only a few hours ago, but this body we're exhuming is centuries old. How can this be? How can this body have proven incorruptible by decay and the ravages of time? According to the Catholic faith, such incorruptibility is a miracle. A person who dies and proves incorruptible can thus qualify as a saint.

There are quite a few alleged examples of this, and we'll take a look at some of the best known. But first, let's examine exactly what the church's criteria are for incorruptibility. In essence it means that the body does not decompose after death, in a miraculous manner not explainable by natural processes. The body has to remain flexible and is supposed to be indistinguishable from sleep; it can't dry out like an Egyptian mummy and be all stiff. The body also must not have been embalmed or otherwise preserved.

The most famous of the Catholic incorruptibles is Saint Bernadette, currently on display at the Chapel of St. Bernadette in France. She died in 1879 and was exhumed thirty years later, so the story goes, and was discovered to be incorrupt and free of odor! However, two doctors swore a statement of their examination of the body, clearly describing a partially mummified corpse, describing the whole body as "shriveled", saying the lower parts of the body had turned black, the nose was "dilated and shrunken", and that the whole body was rigid and "sounded like cardboard when struck." The body was prepared and reburied in a sealed casket. When it was dug up again in 1919, another doctor filed the following report:

> *The body is practically mummified, covered with patches of mildew and quite a notable layer of salts, which appear to be calcium salts... The skin has disappeared in some places, but it is still present on most parts of the body.*

At her third and final exhumation in 1925, it was noted that the "blackish tinge to the face and the sunken eyes and nose would make an unpleasant impression on the public," and so the decision was made to display the corpse with a wax mask. That's right, the photos you see on the Internet of St. Bernadette's beautiful, incorrupt corpse are of a wax mask placed on an obviously mummified body. The descriptions of her condition openly violate all the requirements of incorruptibility, and yet St. Bernadette is the most often cited example of miraculous incorruptibility. When you think about it, if a saint dies and God decides that this body should be incorruptible, you'd think it should remain absolutely perfect, like Sleeping Beauty. It shouldn't be only slightly less decomposed than the average body, and certainly shouldn't be a common mummy.

St. Catherine of Bologna is another nun whose supposedly incorrupt body is still on display. She died in 1463, and although I couldn't find any documentation at all pertaining to the circumstances of her burial and exhumation, the story goes that she was buried without a coffin and was exhumed only 18 days later due to a strong sweet scent coming from her grave — more about that in a moment. Her body is displayed at the chapel of Poor Clares in Bologna, Italy, in a seated position inside a glass case. As you can see from the many photos on the Internet, the body is completely mummified, black and shriveled, and can by no definition be called incorrupt. And yet she is called just that anyway, in utter defiance of the blatantly obvious.

Saint Silvan was a young man said to have been killed for a his faith in the year 350, and his body is on display in Dubrovnik, Croatia, replete with a fresh-looking gash on his throat said to have been the cause of death. The body appears to be perfect. It is a sculpted effigy — St. Silvan's actual remains are said to be contained within the box below the effigy. But there is no display signage to explain this to the faithful, and many come away with photographs of what they think is the actual body. If he is incorrupt as the church claims, why display the effigy instead of the body?

Padre Pio, the 20th century priest famous for his stigmata, is also on the church's list of incorruptibles. However, according to the church's own records, his body was embalmed with formaldehyde upon death. Even so, at his exhumation 40 years later, the remains were described as "partially skeletal" and morticians were unable to restore the face to a viewable condition, so Padre Pio is displayed today with a lifelike silicone mask.

Incorruptible bodies, when exhumed, are often said to be accompanied by a sweet odor which Catholics called the Odor of Sanctity. This odor is also said to come from stigmata on living saints. Some saints are said to have exuded this odor after death. Of course the obvious explanation for such a smell would be embalming fluid. However, modern embalming fluids, basically formaldehyde mixtures, are said to have a strong, unpleasant smell like gasoline. Therefore most manufacturers mask the smell with perfume additives. Historically, sweet-smelling ointments were used on corpses to counter the smell of decomposition, and many such ointments are now known to have contained guaiacol, an effective preservative made from beechwood tar, similar to creosote. So, dig up a body and find it in any state of preservation, and you're likely to smell a strong sweet odor. Evidence of embalming or odor-masking is a better explanation for this smell than some supernatural "Odor of Sanctity".

The best examples of natural incorruptibility come from the peat bogs of northern Europe. About a thousand individuals have been exhumed from the bogs, where a unique combination of cold conditions and chemical processes preserves the soft tissues. Most of these come from the Celtic iron age, but some are far older; the oldest being Koelbjerg Woman who is

5,500 years old. In bog bodies, peat acid actually dissolves the bones but leaves the soft tissues pliable, like rubber, though stained brown and actually tanned into leather. Technically, these bodies meet the Catholic criteria of incorruptibility far better than any of the dried mummified corpses that the church claims. Why are the bog people not considered saints? At least their bodies actually do remain flexible. The church probably says that the natural chemical process counts as embalming, which of course it does; but at least this is a natural process and not a deliberate human embalming as has happened with so many of the so-called miraculous Catholic incorruptibles.

Most recent is the case of Hambo Lama Itigelov, a Buddhist monk who died in the Russian Mongol territory of Buryatia in 1927 and was exhumed in 2002, by his own last request. His condition was described by the monks and a pathologist in attendance as that of someone who had died only 36 hours before. Video shows what looks to be a well-preserved mummy, but hardly that of someone who died only 36 hours before. His body is on display in the open air and is claimed to remain pliable, a claim which is untested. Despite a Russian documentary movie finding no explanation, and the monks' claims to the contrary, the pathologist's own report found the body to have been preserved with bromide salts. Itigelov had also instructed that his body be packed in salt, another way to help prevent decomposition by absorbing moisture away from the body. It's interesting that in life, Itigelov actually had a degree in medicine, and had written a Buddhist encyclopedia on pharmacology.

Buddhist monks have long practiced self-mummification. Some Japanese monks used to prepare themselves for self-mummification through a technique called *sokushinbutsu*. They ate a subsistence diet of nuts and seeds for 1000 days to get rid of all their fat, and then spent the next 1000 days eating only bark, roots, and drinking the tea of a poisonous tree called the urushi, in an effort to make their body both dehydrated and toxic to parasites. Finally they would place themselves inside a stone tomb, ringing a bell once each day. When the bell failed to ring, the other monks would seal the tomb, wait another 1000 days, and then open it up to find out whether the monk had mummified. Only about 20 such monks were successfully mummified in this manner; the rest decomposed normally. Even this number is impressive given that the internal organs remained, which are a prime source of bacteria that contribute to decomposition. Tests of the mummies have revealed toxic levels of arsenic, which is another embalming agent. Together with the lack of body fat and pre-existing dehydration, monks practicing *sokushinbutsu* actually had a reasonable chance of mummification if their tomb was well sealed and conditions were dry. Hambo Lama Itigelov's own technique of using bromide salts and salt packing appears to be a scientifically updated form of *sokushinbutsu*.

But you couldn't call these monks incorruptible any more than you can use the term to describe the Catholic saints who are obviously mummified. Mummification is the natural, expected process that happens to a body under the right conditions. There's nothing miraculous about a natural, expected process. I suppose some people claim that in some of these cases, decomposition should have taken place instead of mummification, and thus the miracle. So, what; leaving a few strands of beef jerky stretched over the bones is the best that the miracle-creating superbeing was able to muster? I'm not convinced, and a skeptical Catholic shouldn't be either. Incorruptible should mean incorruptible. The corpse needs to be flexible and lifelike, as if asleep. We've never seen anything remotely like that. There are no verifiable, viewable examples of supernatural incorruptibility anywhere on the planet, and no reason to think there ever have been.

References & Further Reading

Aufderheide Arthur C. *The Scientific Study of Mummies.* Cambridge: Cambridge University Press, 2003. 273-276.

Edwards, Harry. "Incorruptibility: Miracle or Myth?" *Incorruptibility: Miracle or Myth?* Investigator Magazine, 1 Nov. 1995. Web. 10 Jan. 2010. <http://users.adam.com.au/bstett/PaIncorruptibility.htm>

Faure, Bernard. *The Rhetoric of Immediacy.* Princeton, New Jersey: Princeton University Press, 1991. 148-178.

Nickell, Joe. *Looking for a Miracle: Weeping Icons, Relics, Stigmata, Visions & Healing Cures.* Amherst, New York: Prometheus Books, 1993. 85-92.

Spindler, K., Wilfing, H., Rastbichler-Zissernig, E., Nothdurfter, H. *Human Mummies.* New York: Springer-Verlag Wien, 1996. 161-171.

13. The Oak Island Money Pit

It began as every boy's dream adventure, like a chapter from Tom Sawyer. It was the year 1795 when young Daniel McGinnis, a lad of 16, rowed to Oak Island in Nova Scotia on a journey of exploration. On the eastern end of the wooded island he found something out of place: an old wooden tackle block suspended from a heavy branch, and on the ground below, a sunken depression.

It didn't take young Daniel long to bring in heavy equipment, in the persons of two friends armed with shovels and the knowledge of old stories that the pirate Captain Kidd may have buried treasure on this part of the coast. Two feet down they struck a layer of flagstones, and all the way down they found pick marks on the walls of the shaft. The three boys dug for days, and just when they were about to give up, they came to a solid platform of logs. There was nothing under the logs, but it fired the boys up: There was no longer any question of whether something had been buried here. Over the coming weeks they finally reached 30 feet — that's incredible for three teenagers — and along the way found two more log platforms. By then the difficulty and frustration won, and they gave up.

But the local newspapers had made the pit something of a phenomenon. A few random people came and tried to dig further, and around 1803 a mining group called the Onslow Company took over the island and made the first really serious effort. They found more regularly spaced log platforms, a few of which they found to have been apparently sealed using coconut fiber and putty. Their most important find was a stone tablet at 90 feet, inscribed with weird symbols, translated to mean "40 feet below, two millions pounds lie buried." Encouraged, they continued their dig, but got no further: Just below the tablet they struck a side tunnel, open to the sea, which immediately flooded the pit. It was a booby trap.

To make a very long story short, many companies and investor groups have taken over the island and launched major digging efforts, costing millions of dollars and the lives of six men killed in various accidents. An auger sent down the hole in 1849 past the flood tunnel went through what was said to be a sheet of iron, more oak, "broken bits of metal", and brought up three links of gold chain and a 1-centimeter scrap of parchment reading "vi" or "ri". Flooding and collapses marred many mining efforts over the decades. In 1965 a causeway was built to the island to deliver a great 70-ton digging crane, which excavated the Money Pit to a depth of 134 feet and a width of 100 feet. In 1971, workers sunk a steel caisson all the way down, finally striking bedrock at 235 feet. They lowered a video camera down into a cavern at the bottom of the shaft, but whether it shows nothing interesting or a spectacular pirate's cave seems to depend on whose account you read. After more than 200 years of excavation, that entire part of Oak Island is now a wasteland of tailings and abandoned gear, and the location of the original Money Pit is no longer known.

What might have been buried at the Money Pit? Theories abound beyond the popular local notion of Captain Kidd. Other pirates said to have buried loot in the region include Henry Morgan and Edward Teach, as well as Sir Francis Drake and Sir Francis Bacon. Some suggest the Vikings. An interesting nomination is that of the British Army protecting payroll during the American Revolution, whose Royal Corps of Engineers were about the only folks at that time with the ability to construct such a system, with subterranean flood tunnels leading to the sea.

I've long been curious about a couple of elements from the Oak Island story. First, the coconut fiber. Coconuts are not found in Nova Scotia, nor anywhere in the vicinity. The closest place coconuts were found in the 18th century was Bermuda, about 835 miles due south of Oak Island. You might think this supports the pirate theory, since pirates certainly frequented Bermuda and the Caribbean. Some accounts of Oak Island have said that coconut fibers were found in large quantities buried beneath the beach, though this has never been evidenced. A 1970 analysis by the National Research Council of Canada did identify three of four samples submitted as being coconut fiber. Radiocarbon dating found that the coconut came from approximately the year 1200 — three centuries before the first European explorers visited the region, and two centuries after the only known Viking settlement more than 600 miles away. How would 300-year-old coconuts get buried 50 feet underground when there was nobody around to do it, in a booby-trapped shaft that nobody had the technology to dig?

It's really those flood tunnels that put the Money Pit over the top of anything you could expect pirates or anyone else to be responsible for, especially during a century when no Europeans were within thousands of miles. Divers would have needed weeks or months to cut subterranean tunnels from the bay nearby to the 90-foot level of the Money Pit. Or would they?

The geology of Oak Island and its surrounding area gives us some more clues. The region is primarily limestone and anhydrite, the conditions in which natural caves are usually formed. In 1878, a farmer was plowing Oak Island just 120 yards away from the Money Pit when suddenly her oxen actually broke through the ground, into a 12 foot deep sinkhole above a small natural limestone cavern. 75 years later, just across the bay, workers digging a well encountered a layer of flagstone at two feet, and as they dug to a depth of 85 feet, they encountered occasional layers of spruce and oak logs. Excitement raged that a second Money Pit had been found, but experts concluded that it was merely a natural sinkhole. Over the centuries sinkholes occasionally open up, trees fall in, and storms fill them with debris like logs or coconuts traveling the ocean currents. These events, coupled with the underground cavern at the bottom of the Money Pit discovered in 1971 and the discoveries of numerous additional sinkholes in the surrounding area, tell us that Oak Island is naturally honeycombed with subterranean limestone caverns and tunnels. The geological fact is that no Royal Corps of Engineers is needed to explain how a tunnel open to the sea would flood a 90 foot deep shaft on Oak Island, booby trap style.

Obviously the story has plenty of elements not thoroughly explained by the theory that the Money Pit was simply a natural sinkhole consistent with others in the area. One such element is the stone tablet. No photographs exist of the stone, nor any documentation of where it might ever have physically been. The transcribed markings are in a simple substitution cipher using symbols borrowed from common Freemasonry, and they do indeed decode as "Forty feet below, two million pounds lie buried", in plain English. The stone tablet made its appearance in the Onslow Company's records, coincidentally, about the same time they were running out of money and their pit flooded. Most researchers have concluded that the stone tablet was probably a hoax by the Onslow Company to attract additional investment to continue their operation. The same can be said of the other two significant artifacts, the links of gold chain and the parchment. Accompanying his map, Joe Nickell said "The artifacts are not properly documented archaeologically, and most would appear to derive from historical activity on the island or from subsequent excavation or hoaxing by workmen."

And as with so many other subjects, the older the account you read, the less specific and impressive the details. The contemporary newspaper accounts of Daniel McGinnis and his two friends make no mention of a tackle block or of regularly spaced log platforms, only that logs were found in the pit, and that the tree branch showed evidence of a block and tackle having been used. Armed with proper skepticism and the willingness to look deeper than the modern sensationalized retellings, the Money Pit's intrigue and enchantment begin to fade.

I was probably no more than eight or nine when I first read about young Daniel McGinnis and his treasure tree, and at that very moment, Oak Island became a permanent part of me, as it has so many others. Oak Island's history is a patchwork of individual romances and adventures, a tapestry made from the reveries of skeptics and believers alike. Whether building causeways and sinking caissons, analyzing old newspapers, swinging a pick in the glare of a lantern, or even listening to a podcast about the mystery, all of us share the same ambition. No matter if we seek treasure or truth, we all long for the chance to turn just a few thrusts of the shovel, and we care not what we find.

References & Further Reading

Joltes, R.E. "A Critical Analysis of the Oak Island Legend." *History, Hoax, and Hype: The Oak Island Legend.* Critical Enquiry, 19 Apr. 1996. Web. 14 Jan. 2010.
<http://www.criticalenquiry.org/oakisland/analysis.shtml>

Lamb, L. *Oak Island Obsession: The Restall Story.* Toronto: Dundurn Press, 2006. 196.

Nickell, J. "The Secrets of Oak Island." *Skeptical Inquirer.* 1 Mar. 2000, Volume 24, Number 2: 14-19.

O'Connor, D. *The Secret Treasure of Oak Island: The Amazing True Story of a Centuries-Old Treasure Hunt.* Guilford: The Lyons Press, 2004.

Teale, E. "New clews aid search for Oak Island's gold." *Popular Science.* 1 Jun. 1939, Volume 134, Number 6: 106-109, 226, 228.

14. Space Properties for Sale

Get out your credit card and start snapping up property throughout the cosmos: The Internet is stuffed with companies claiming the right to sell you a piece of extraterrestrial real estate or to name some celestial body. Own an acre on the Moon, found your own ski resort on the slopes of Olympus Mons, name a star for your sweetheart. It's all out there, waiting for the next sucker to lighten his wallet.

The International Star Registry was the first of these companies, so far as I've been able to find. It began in 1979 with ads in the backs of magazines, and would mail you a certificate describing the location of your star and giving every indication that that star was now officially named after you or whoever you paid for it to be named after. With its official sounding name, the International Star Registry's official status was rarely questioned. In the ensuing three decades, copycats have sprung up everywhere. LunarEmbassy.com sells not only property on the moon, but also "lunar corporations" for only $379, and boasts an impressive celebrity customer list. MarsShop.com does the same thing for Martian real estate. AsteroidDatabase.com lets you give any asteroid the name of your choice. And a whole ridiculous host of knockoffs have appeared for naming stars: BuyTheStars.com, UniversalStarCouncil.com, NameAStar-Live.com, and countless others. They offer different packages. They promise to donate to charity. They speak impressively of their plans to develop said properties. They promise to publish your ownership in a book, or to place their copyrighted list into the Library of Congress.

None of them ever seem to get around to mentioning how they managed to acquire the right to sell these properties.

That's because none of them have any such rights. They can't sell you property on the Moon because they don't own it. And neither the International Star Registry nor any of its imitators have any more right to name stars than your cat does. What they *can* do, and what some of them even honestly admit if you can find it buried on their disclaimer page, is that they're merely selling you a slot in some unofficial, unrecognized list of their own, which is perfectly legal; often calling it a "novelty gift". Sometimes these companies use deceptive or vague language hoping you will infer that the named or purchased property is legitimately recognized. For example, the International Star Registry used to say on its web site:

> *Because these star names are copyrighted with their telescopic coordinates in the book, Your Place in the Cosmos, future generations may identify the star name in the registry.*

Hey, it's copyrighted, sounds legitimate. But in 2006, perhaps due to pressure from consumer groups and the astronomical community, they removed that statement and posted a new FAQ page that gave a much more honest and straightforward answer:

> *Q: Will the scientific community recognize my star name?*
> *A: No. We are a private company that provides Gift Packages. Astronomers will not recognize your name because your name is published only in our Star catalog.*

Dennis Hope, the founder of LunarEmbassy.com, breaks with form and asserts legal ownership of the Moon and most of the other planets and moons in the solar system. He states that he simply "claimed" ownership and that such a claim is valid. The legal foundation comes from the 1967 Outer Space Treaty, to which all spacefaring and most other nations are signatory. It states in part that no nation can own any part of the Moon or other planets and moons. Dennis Hope states that it says nothing about individuals and corporations owning them, so he feels this loophole allows his claim to thus grant him full ownership of the real estate in the solar system. But experts on space law don't give Hope's thoughts on the matter too much credence. Ram Jakhu, law professor at the Institute of Air and Space Law at McGill University in Montreal, points out that "Individuals' rights cannot prevail over the rights and obligations of a state." Hope's "loophole" is already thoroughly closed by other statutes well precedented. Thus, Hope's claim is a little bit like saying it's OK to park your car illegally because the speeding laws say nothing about illegal parking.

There is actually one, and only one, governing body with the authority to manage the official names of celestial bodies, the International Astronomical Union. In response to all the numerous requests they've received over the years to name stars or asteroids or to inquire about the validity of companies offering space properties for sale, the IAU published a statement. It says in part:

> *Some commercial enterprises purport to offer such services for a fee. However, such "names" have no formal or official validity whatever... Similar rules on "buying" names apply to star clusters and galaxies as well. For bodies in the Solar System, special procedures for assigning official names apply, but in no case are commercial transactions involved.*

All known stars are already named, according to various conventions, usually with an alphanumeric designation. Various methods of cataloging stars are used, such that most stars are listed redundantly in multiple catalogs. Designations will often incorporate the star's position in the sky either by its constellation or by the angle of its position; they may incorporate its brightness, its type, or a serial number. Most well known stars also have common names, like Deneb, Regulus, Vega, or Sirius. Beyond these established conventions, any name that anyone might make up and sell for a star has no astronomical legitimacy or permanence outside the records of the company who sold it.

Asteroids, or "minor planets", are different from stars in that they can be named. But not just anyone can name them. First of all, you have to be the discoverer. You have to submit your discovery to the Minor Planet Center, which, if your discovery is confirmed, will assign a designation based on the year of discovery, two letters encoding the date of discovery, and a serial number. This designation only becomes permanent if the orbit of the minor planet can be accurately enough described that its future position can be predicted. If this happens, the Minor Planet Center invites the discoverer to submit a name for the object, which must conform to a lengthy set of specific rules. A 15-person committee at the Minor Planet Center then confirms or denies the submitted name. If the discoverer does not submit a suitable name before the ten years expire, the right to name it falls to the Minor Planet Center. At no time is it possible for any commercial transaction to transfer the right to name a minor planet.

The IAU also gives a tidbit of advice for anyone contemplating a purchase of real estate from LunarEmbassy.com or any other extraterrestrial real estate seller:

> *Chances are that [your lawyers] will either laugh their heads off or politely suggest that you could invest their fees more productively. At a minimum, we suggest that you defer payment until you can take possession of your property.*

It seems incredible that it has to be repeated, but the fact is that just because someone is selling something on the Internet doesn't make it worthwhile, or mean that it even exists at all. The right to legitimately name or own space properties does not exist outside of the novelty gift industry, and is not worthy of your checkbook's attention. Don't be fooled; be skeptical.

References & Further Reading

Andersen, Johannes. "Buying Stars and Star Names." *International Astronomical Union*. International Astronomical Union, 1 Jan. 1999. Web. 3 Jan. 2010. <http://www.iau.org/public_press/themes/buying_star_names/>

Berman, Bob. *Strange Universe, the weird and wild science of everyday life--on Earth and beyond.* New York: Henry Holt and Company, 2003. 158-161.

Britt, Robert Roy. "Name a Star? The Truth about Buying Your Place in Heaven." *Nightsky*. Space.com, 15 Sep. 2003. Web. 3 Jan. 2010. <http://www.space.com/spacewatch/mystery_monday_030915.html>

Kidger, Mark. *Astronomical Enigmas, Life on Mars, the star of Bethlehem & other milky way mysteries.* Baltimore: The Johns Hopkins University Press, 2005. 46-47.

Plait, Philip. *Bad Astronomy, Misconceptions and misuses revealed, from astrology to the moon landing 'hoax'.* New York: John Wiley & Sons, 2002. 236-244.

Wuorio, Jeff. *How to buy & sell just about everything, more than 550 step-by-step instructions for everything from buying life insurance to selling your screenplay to choosing a thoroughbred racehorse.* New York: The Free Press, 2003. 95-96.

15. The Bohemian Club Conspiracy

Today we're going to point our skeptical eye at some of the stories surrounding San Francisco's Bohemian Club. The membership rolls of this exclusive, men-only social club include US Presidents, captains of industry and banking, military contractors, giants in the arts and sciences, and the absurdly wealthy. Thus, the Bohemians are automatically the subject of charges by some that they are a secret guild of Illuminati, making the decisions that rule the world; manipulating interest rates and choosing political candidates, for example. But from that point the Bohemians deviate from other usual targets of such conspiracy theories, with their unique annual encampment at the Bohemian Grove where they are said to participate in some most un-Illuminati-like behavior, up to and including mock human sacrifice to the pagan god Moloch.

The bare facts are that the Bohemian Club was founded in San Francisco during the gold rush, by some of the social elite in the city's arts community. By mixing wealthy patrons with artistic luminaries, they ensured that the arts would continue to flourish, while providing patrons and enthusiasts with the ultimate hobnobbing experience. Their motto *Weaving spiders come not here*, taken from Shakespeare's *A Midsummer Night's Dream* and meaning that no business is to be transacted at the club, was cast in bronze and is prominently displayed on their six story building.

Once a year, Bohemian Club members (often called Bohos) retire to the Bohemian Grove, their 2700-acre retreat in the redwoods of Sonoma County, north of San Francisco, for 16 days, separated into fraternity-style camps. It is at this annual encampment that their most controversial activities are said to take place; everything from bizarre pagan and homosexual rituals, to plans for global domination. The Sonoma County Airport fills with a staggering collection of corporate and military business jets. Grove security is said to be exceptionally tight. The *Sonoma County Free Press* maintains the Bohemian Grove Action Network to organize round-the-clock protests during the camp.

Conspiracy theorists love the Bohemians. One, a 9/11 conspiracy theory radio host named Alex Jones, speaks extensively of them, sells self-published books and DVDs "exposing" their evil doings, and posts exhaustive YouTube videos (here and here) on the subject. I've found his claims to be fairly lightweight, like "world leaders meet at the Grove". Well, so they do. Some charge that it's immoral for powerful people to congregate, saying that when powerful people work together, they become even more powerful. Some claim that the "Lakeside Talks" contain confidential information, or that the world's military-industrial complex is driven by Bohemian committees.

So I decided I wanted to know how much truth there was to these rumors about the Bohemians, if any. The first question to ask is: Can you know? Being an outsider and having to rely on unofficial reports, can you trust that the information you hear is reliable? Would this be like trying to find out what happens in the inner sanctum of a Mormon Temple, or at the highest levels of Scientology, where the only stories are coming from disgruntled former members?

Following the scientific method, we start by recognizing that we can't prove a negative. There's no way we can certify that nothing evil ever happens at the Bohemian Club. What we can do is find evidence of any nefarious activities. For example, anti-Bohemian web pages routinely make vague claims like who our presidential candidates are going to be is decided at the Bohemian Grove. Well, what's the evidence? Do we have tapes of this? Was there a witness? Or is it just pure speculation?

It's said that Nixon and Bush 43 both announced their running mates at the Bohemian Grove. Well, so what if they did? It's a fine place to make such an announcement. But that's a far cry from claiming that an evil committee of Bohemian insiders made the decision and *told* Nixon and Bush who their running mates would be. I've no doubt that plenty of talk happens at the Bohemian Grove that also might as well happen at golf courses and restaurants. That's normal human interaction, not a conspiracy to rule the world. To justify conspiracy charges, you can't simply find that the circumstances exist in which it might be convenient to hold a meeting of conspirators. Said meeting must actually take place.

I'm not coming into the Bohemian question totally blind. I've actually been to two events at The Bohemian Club in San Francisco. One was a wedding and not a club sanctioned event, but the other was a Steve Miller Band concert held in their 600-seat theater, followed by an open public jam session in a large ballroom. I heard anecdotally that they were trying to get Steve Miller to join. I saw nothing unusual: Lots of books, portrait art, and a really busy cash bar; but then I was only there as a guest and was not invited into any secret meetings. The one thing I learned for sure is that drunken Bohemians onstage with the Steve Miller Band do not add appreciably to the concert experience.

The Bohemian Grove camp is dominated by such theatrical attractions. Each year ends with what they call the *High Jinks*, an original play written each year for the occasion and performed by a cast of hundreds, and said to be usually quite good and magnificently performed with six-figure budgets.

Throughout the camp, so-called *Low Jinks* are also performed, which can get pretty lowbrow and are the source of the stories about Bohos dressed in drag.

So what about this mock human sacrifice to the pagan god Moloch? It's well documented from reporters and others who have snuck in and photographed the ceremony, which takes place on the first day of the 16-day camp. A 50-minute dramatic play accompanied by a full orchestra, with huge production value, climaxes with figures cloaked in robes burning a coffin on a sacrificial altar before a huge cement owl. Sounds pretty strange to me. This is what our US Presidents are doing when they visit the redwoods?

The Bohemians call this the *Cremation of Care* ceremony, and the name of the antagonist in the coffin is *Dull Care*. It turns out the term dull care comes from a 1919 Oliver Hardy short film of that title, in which the police force is more interested in being lazy than in chasing bad guys. They leave the "dull cares" of their day behind them, and enjoy their freedom. The Bohos thus kick off their encampment by metaphorically throwing their "cares of the day" onto the sacrificial pyre. The theatrics of the robes and the ceremony are certainly consistent with the dramatic theme of the camp. The giant owl merely represents knowledge, and it's been the club's symbol since its inception.

Moloch, also spelled Molech, is the Hebrew word for *king* and was mentioned in the Bible as a Canaanite demigod who required child sacrifices. I couldn't find any classical references in which Moloch is depicted as an owl, rather he has a human body with the head of a bull. As far as I can tell, any connection between the demigod Moloch and the Bohemians is simply made up by people who are so desperately trying to find something to dislike about them. Burning things, whether it's witches, festival effigies, dull cares, or magnesium engine blocks happens in all cultures all around the world without the assistance of Canaanite demigods. I'm left with the null hypothesis here: That the *Cremation of Care* ceremony is nothing more than the Bohos say it is. Interesting how easy it is to make something sound weird when you omit details like the weird cloaks are the costumes of characters in a play, not the ceremonial garb of Bohemian elders.

It turns out that the conspiracy story most often told about the Bohemians is actually true: That the Manhattan Project was planned there, to develop the first atomic weapons. The Bohemian Grove property is vacant 49 weeks out of the year, and when it is, it's available for private use by members. In September of 1942, Dr. Edward Teller, a club member, reserved the grove and held his meeting there. No other Bohemian Club members were present. So although it's factually true that the Manhattan Project was planned at the Bohemian Grove, the obvious implication that it was a Bohemian Club project planned by Bohemian Club members is patently false.

So you've got to wonder, if there's nothing evil going on, what's it's really all about? Why do these wealthy, powerful men go into the woods and act like drunken frat boys from the drama club? If they're not there to plan global domination, why bother? These are men whose time is worth something; they're hardly going to go screw around at the Bohemian Grove without a good reason. William Domhoff from the Department of Sociology at UC Santa Cruz came up with five reasons in his doctoral dissertation studying the behavior of the Bohemians:

1. Physical proximity is likely to lead to group solidarity.

2. The more people interact, the more they come to like each other.

3. Groups seen as high in status are more cohesive.

4. The best atmosphere for increasing group cohesiveness is one that is relaxed and cooperative.

And to explain why these first four are useful:

5. Social cohesion is helpful in reaching agreement when issues are introduced into experimental groups for resolution.

Basically, it's a giant team building exercise. Think back to the club motto: *Weaving spiders come not here.* The Bohemian Club is not a place to do business; it's a place to build and strengthen the relationships you'll need when it comes time to do business out in the real world. The suggestion that the Bohemian Grove is really just a large-scale team building exercise is supported by the fact that their security is not anything like what it's popularly rumored to be. In 1989, Philip Weiss, writing for *Spy* magazine, strolled on into the Grove and reported the following:

> *The sociologists who had studied the place were right; there was no real security... I took care to blend. I outfitted myself in conservative recreational wear... I always carried a drink, and I made it a point to have that morning's Wall Street Journal or New York Times under my arm when I surfaced... Thus equipped, I came and went over 7 days during the 16-day encampment, openly trespassing in what is regarded as an impermeable enclave and which the press routinely refers to as a heavily guarded area. Though I regularly violated Grove rule 20 ("Members and guests shall sign the register when arriving at or departing from the Grove"), I was never stopped or questioned... Indeed, I was able to enjoy most pleasures of the Grove, notably the speeches, songs, elaborate drag shows, endless toasts, pre-breakfast gin fizzes... and other drinks -- though I didn't sleep in any of the camps or swim naked with likeminded Bohemians in the Russian River at night.*

I conclude that conspiracy charges against the Bohemian Club are vague, implausible, and without any supporting evidence. The null hypothesis, that it is simply a team-building event among wealthy and powerful men interested in the arts and otherwise, is well evidenced and consistent with everything that's publicly known about them. Of those who say there's more to it, I'm going to remain skeptical.

References & Further Reading

Aaronovitch, D. *Voodoo History: The Role of the Conspiracy Theory in Shaping Modern History*. New York: Riverhead, 2010.

Domhoff, G. William. *The Bohemian Grove and Other Retreats*. NY: Harper and Row, 1974. 1-80, Appendix.

Phillips, Peter M. *A Relative Advantage: Sociology of the San Francisco Bohemian Club*. Davis, CA: UC Davis, 1994.

Ronson, Jon. *Them: Adventures with Extremists*. New York: Simon & Schuster, 2002. 295-327.

Sancton, Julian. "A Guide to the Bohemian Grove." *Vanity Fair Magazine*. Vanity Fair, 1 Apr. 2009. Web. 17 Jan. 2010. <http://www.vanityfair.com/style/features/2009/05/bohemian-grove-guide200905>

Sides, Hampton. *Stomping Grounds: A pilgrim's progress through eight American subcultures*. NY: Morrow, 1992.

Weiss, Philip. "Inside Bohemian Grove." *Spy*. 1 Nov. 1989: 59-76.

16. The Sargasso Sea and the Pacific Garbage Patch

Today we're going to sail the oceans blue, through the wind and spray, at least until we become mired in two remote wastelands said to be the bane of mariners: The Sargasso Sea in the middle of the Atlantic, and its companion on the other side of the globe, the Pacific Garbage Patch (also called the Pacific Trash Vortex, and various other names). Both areas are encircled by swirling ocean currents, and are said to collect their debris in the center. Both are the subject of tall tales. Both are ripe for skeptical inquiry.

The Sargasso Sea

Let's start in the Atlantic. The Sargasso Sea is actually a named sea, and it's the only one in the world with no shorelines, being completely enclosed within the North Atlantic Ocean. It's a 2,000,000 square mile oblong oval stretching across the Atlantic, centered on about 30°N latitude, bounded by major clockwise ocean currents. It's best known in stories for being a dense mat of solid seaweed, a tangled mass that no ship can penetrate. You hear tell of ancient mariners finding abandoned wrecks trapped, or starving crews who tried to walk out. Most famously, Christopher Columbus wrote about the Sargasso Sea in his log. Upon encountering the seaweed, he thought he must be near land, but when no land appeared after days of sailing his crew almost mutineed. Jules Verne's *Nautilus* visited the Sargasso in his novel *20,000 Leagues Under the Sea*:

> Such was the region the Nautilus was now visiting, a perfect meadow, a close carpet of seaweed, fucus, and tropical berries, so thick and so compact that the stem of a vessel could hardly tear its way through it. And Captain Nemo, not wishing to

> *entangle his screw in this herbaceous mass, kept some yards beneath the surface of the waves. The name Sargasso comes from the Spanish word "sargazzo" which signifies kelp. This kelp, or berry-plant, is the principal formation of this immense bank... Above us floated products of all kinds, heaped up among these brownish plants; trunks of trees torn from the Andes or the Rocky Mountains, and floated by the Amazon or the Mississippi; numerous wrecks, remains of keels, or ships' bottoms, side-planks stove in, and so weighted with shells and barnacles that they could not again rise to the surface.*

In 1910, the steamer *Michael Sars* took a party of scientists out to study the Sargasso Sea for three months, and their report was among the first that debunked these legends. They found the patches of seaweed to be small and sparse, rarely larger than a doormat, and in no way a hazard to navigation. They also found the waters of the Sargasso Sea to be warmer, clearer, bluer, and with less oxygen than the surrounding Atlantic. For this reason they declared the Sargasso Sea to be a type of desert, largely bereft of sealife.

This notion was challenged in 2007 when research published in the journal *Science* found that phytoplankton blooms were the reason for the decreased oxygen. The rotation of the huge eddy draws up salty, nutrient rich water from the ocean floor, enriching the sunlit upper layers of water, and kick-starting an unusually active ecosystem. Thus the Sargasso "desert" is actually quite the opposite, and the relative proliferation of *sargassum* seaweed is due to nutrient rich growing conditions, not vortex suction action.

What of the tall tales of ships becoming trapped, and stranded sailors dying? It so happens that the Sargasso Sea is a convergence of several conditions, not just the nutrients from the deep, but also weather conditions at its latitude. 30-35° latitude, both North and South, is the location of the subtropical ridge between the trade winds. Conditions here are variable winds and low precipitation. Sailors call it the doldrums. Sailing vessels entering the Sargasso Sea are virtually certain to come to a grinding halt in a dead calm. Throw in some strange seaweed and we have all the ingredients for a nautical epic.

THE PACIFIC GARBAGE PATCH

But if we spin the globe and look at the center of the North Pacific Ocean, we see a phenomenon that *is* due to vortex suction — the Pacific Trash Vortex, also called the Pacific Garbage Patch. One guy emailed me that he'd looked for it on Google Earth; he'd heard there was a giant island of solid floating garbage twice the size of Texas in the middle of the Pacific Ocean.

The North Pacific Gyre is a clockwise rotation of ocean currents in the North Pacific. Wind and currents combine to drive floating matter toward its center. You might have heard of the Pacific Garbage Patch before, and are probably just now wondering why you've never seen any photographs of a giant island of trash. In fact, Hawaii is right in the center of the Gyre, and if there were a Pacific Garbage Patch, Hawaii would be in its exact center. The answer is simple: No such floating island of trash exists. Despite the fact that there is a huge amount of plastic waste in the Pacific Ocean, and despite the fact that the Gyre does drive it all toward the center, there is no floating island. How do we corroborate these two seemingly mutually exclusive facts?

One proposed explanation, put forward in a map created by Greenpeace, shows the garbage patch as two separate patches on either side of Hawaii, both well clear of it; thus nobody ever sees them. The map also gets East and West reversed, and is dramatically wrong in its depiction of ocean currents, splitting the North Pacific into two counter-rotating swirls, instead of one big one like it actually is. So let's set aside Greenpeace's version for the moment, and go back to the origin of the story.

In 1988, Robert Day, David Shaw, and Steven Ignell submitted a report to NOAA (National Oceanic and Atmospheric Administration) detailing the results of four years of sample collection and analysis of plastic fragments found floating in the Pacific Ocean. They found concentrations highest in the North Pacific Gyre. The authors cited wind and currents as the primary force driving the higher concentrations to the center of the Gyre. Concentrations of what? Number one, monofilament fishing line fragments; and number two, something called neuston plastic. Neuston plastic refers to particles that have been broken down to a small size and are now floating just at or below the surface of the water. Most plastic floating in the open ocean degrades quite quickly, due primarily to ultraviolet radiation. It becomes brittle and crumbles. When it reaches microscopic size, it competes with phytoplankton as a food source for zooplankton, and enters the food chain. That's not good for anyone. The authors used 203 sample stations, each about 450 square meters in size. 52.2% of these contained plastic fragments.

Got that? Only *half* of NOAA's football-field sized sample areas, in the center of the densest part of the Pacific "Garbage Patch", contain even detectable levels of microscopic plastic. Unacceptable to be sure; but hardly a solid island.

The reason is that getting to the center of the Gyre takes more than enough time for plastic to break down. Oceanographer and sailor Charles Moore estimates that garbage from Asia takes about one year to reach the Gyre, and about five years from the United States. Moore is largely responsible for bringing the issue into the public eye. Upon his return voyage from the 1997 Transpac ocean race from California to Hawaii, he wrote:

> *There were shampoo caps and soap bottles and plastic bags and fishing floats as far as I could see. Here I was in the middle of the ocean, and there was nowhere I could go to avoid the plastic.*

Although Moore is doing important work, some of his more overly dramatic descriptions like this one have helped to launch the popular belief in Texas sized garbage patches.

Bringing attention to the issue is good; presenting an over-dramatized representation of the facts to do so, not so much.

Now, I'm not here to defend the dumping of trash at sea, which is the default criticism I'm going to receive for pointing out that there is no Texas-sized island of trash surrounding Hawaii. I remember once while sailing from Newport Beach to Cabo San Lucas, about 100 miles offshore we crossed the path of a cruise ship that had passed in the night. It actually left a visible path: mainly an oil slick, dotted with party balloons, plastic cups, and other junk. Very nice. I also remember the first time I saw a garbage scow being towed out to sea, loaded with an acre of trash, to be dumped. I grabbed a marine chart and saw there was actually a marked area offshore designated for such dumping. I couldn't believe it. So while I do have opinions on the subject, my role here is to focus on the truth of the stories about a huge solid island of trash floating in the middle of the Pacific. And the truth is there isn't one.

REFERENCES & FURTHER READING

Berlofff, Pavel S. et. al. "Material Transport in Oceanic Gyres." *Journal of Physical Oceanography.* 10 Jun. 2001, Volume 32: 764-796.

Kubota, M. "A Mechanism for the Accumulation of Floating Marine Debris North of Hawaii." *Journal of Physical Oceanography.* 1 May 1994, Volume 24, Issue 5: Pages 1059–1064.

LiveScience Staff. "Mystery of the Sargasso Sea Solved." *LiveScience.* Tech Media Network, 22 May 2007. Web. 11 Jan. 2010. <http://www.livescience.com/strangenews/070522_sargasso_sea.html>

McGillicuddy, D.J. Jr and A. R. Robinson. "Eddy-induced Nutrient Supply and New Production in the Sargasso Sea." *Deep Sea Research Part I: Oceanographic Research Papers.* 1 Aug. 1997, Volume 44, Issue 8: Pages 1427-1450.

NOAA. "Great Pacific Garbage Patch." *NOAA Marine Debris Program.* National Oceanic and Atmospheric Administration, 1 Sep. 2009. Web. 11 Jan. 2010. <http://marinedebris.noaa.gov/info/pdf/patch.pdf>

17. Chasing the Min Min Light

It's a hot night out here in the Australian outback. We're about 40 miles east of the tiny village of Boulia, near the site of an old ghost town called Min Min that doesn't even exist anymore. They call this part of western Queensland the channel country, a flat expanse of flood plains and creeks and dry washes that almost never flow. Stand on a high point and you can see to the horizon in every direction. And if you're lucky, on a night like this, you might see the region's most infamous denizen: The Min Min Light.

Sometimes it's an orange speck, sometimes it's a big white ball. Sometimes it's close and sometimes it follows you, sometimes it just hovers away out in the distance. Some say it's just a tall tale, and others say they've seen it with their own eyes. It's been called a ghost light, a spirit orb, and in later years, all manner of scientific sounding natural phenomena. But it's real and it's there, and only a lucky few will see it, but see it they do, and have for the better part of two centuries. We can go all the way back to 1838 to find the earliest known written account of the Min Min Light. It comes from T. Horton James' book, *Six Months in South Australia:*

> *A group of explorers were camped in the Ovens River region of eastern Victoria, when they saw a fire a little way off. Some of them rode off to investigate, but it was about three hours before they returned, and had seen neither fire, bushrangers, nor travelers. They rode boldly up to the spot where the fire, as they thought, was burning, but it was as far off as when they started. In short, it turned out to be an ignis fatuus, or jack-a-lantern, and kept them upwards of an hour trotting on in the vain pursuit, 'till by some sudden flickering and paleness, it confirmed them in its unsubstantial nature, and they returned rather mortified to bed...*

It was a full century before the best known Min Min Light encounter was published in a 1937 issue of Australia's late *Walkabout* magazine. Rancher Henry Lamond told of an adventure that happened to him 25 years earlier:

> ...*I did not leave the head station until about 2 am, expecting to get to Slasher's well before daylight... 8 or 10 miles out on the downs I saw the headlight of a car coming straight for me. Cars, though they were not common, were not rare. I took note of the thing, singing and trotting as I rode, and I even estimated the strength of the approaching light by the way it picked out individual hairs in the mare's mane.*
>
> *Suddenly I realized it was not a car light — it remained in one bulbous ball instead of dividing into the 2 headlights, which it should have done as it came closer; it was too green-glary for an acetylene light; it floated too high for any car; there was something eerie about it.*
>
> *The light came on, floating as airily as a bubble, moving with comparative slowness ... I should estimate now that it was moving at about 10 mph and anything from 5 to 10 feet above the ground ... Its size, I would say, at an approximate guess, would be about that of a new-risen moon. That light and I passed each other, going in opposite directions. I kept an eye on it while it was passing, and I'd say it was about 200 yards off when suddenly it just faded and died away. It did not go out with a snap -its vanishing was more like the gradual fading of the wires in an electric bulb.*

As long as people have been witnessing the light, they've been trying to figure out what it is. Author Mark Moravec came up with five possible explanations for the Min Min Light in his article *Strange Illuminations: Min Min Lights: Australian Ghost Light Stories*:

- Misidentifications of natural phenomena such as wind-blown mists; phosphorescence in marshes; spontaneous neuronal discharges in the visual field; clusters of luminescent insects; light refraction effects; ball lightning or other electric discharge.

- An unknown natural phenomenon involving low-level air oscillations; or ionization in geophysically generated electrical fields.
- Psychokinetic or poltergeist effects unconsciously produced by an individual.
- Non-physical apparitions/ghosts.
- Small, physical UFOs ("remote-control probes").

These nominations are all quite fanciful, and would be difficult to test. But they're not without historical basis. In a 1955 issue of *Walkabout*, writer Ernestine Hill found plausible cause for a ghost light in the history of Min Min:

> *So many were its crimes and murders of kerosene and brimstone, that in righteous anger they burnt it to the ground. The place was stories and desolation- but the dead men would not be forgotten on their stoney plain. Just as a rider was passing by, out of that graveyard came the biggest Jack-O'-Lantern in Australia!*

Most intriguing to the scientific mind, though, are the suggestions that some interesting natural wonder is responsible for the lights. The Marfa Lights in Texas are a similar phenomenon. Piezoelectric effects have been proposed as an explanation for each. The piezoelectric effect is observed in certain crystals that change shape when an electrical current is applied to them, making them useful for such applications as the tiny sound element in a beeping digital watch. The effect also works in reverse: Apply mechanical force to the crystal and it produces an electrical current. A few naturally occurring minerals exhibit this effect, namely some types of quartz, topaz, and tourmaline structures. The idea is that geologic forces within the rock, either tectonic or thermal, create these piezoelectric charges. But there have always been two problems with this hypothesis: First, the effect is a weak electrical voltage, not light; and second, the voltage is measurable only on the crystal itself, it is not projected up into the air in the form of a glowing orb. At least, no such effect has ever been observed or predicted by calculation.

However, an article published in the journal *Nature* in 2006 by authors Eddingsaas and Suslick described a newly discovered form of mechanoluminescence, light from mechanical force. Scientists have known about mechanoluminescence for centuries, since it was first observed as sparks when scraping lumps of sugar. Eddingsaas and Suslick found that shockwaves of acoustic cavitation in a mechanically aggravated crystal slurry produced light much brighter than mechanical crushing alone. Very cool, but also a poor explanation for the Min Min Light. Neither crystal slurries nor sufficient mechanical aggravation are present at the Min Min site, and even if they were, the light is produced immediately at the source of the acoustic cavitation, and not projected up into the air in the form of a floating ball.

Optical science has produced a far better candidate for our mystery ghost light, one that is reproducible and that fully accounts for the observations. Professor Jack Pettigrew, writing in the journal *Clinical and Experimental Optometry*, described *The Min Min Light and the Fata Morgana*. Named for King Arthur's sister Morgana, who was said to be able to levitate objects, a Fata Morgana mirage is one type of superior mirage. In a superior mirage, the object is seen above its actual position, for example above the horizon, when in fact the object is far away and hidden below the horizon. Fata Morgana mirages are caused by thermal inversion layers in the atmosphere. They are fairly common in the arctic regions where temperature inversions are endemic. Often the ocean water temperature, hovering right around freezing, is warmer than the air directly above it, which can be well below freezing. A line of sight stretching along this gradient toward the horizon can easily become distorted, bending light like a fiber optic cable. Arctic islands that are low lying and below the horizon are often seen floating in the air above the horizon, inverted or doubled or stretched into tall blocky mesas.

Pettigrew found that Australia's channel country naturally creates ideal conditions for superior mirages. All of the hollows and ravines trap warm air, and on a cool evening following a hot day, the conditions are virtually certain to take place. Pettigrew and six observers parked a car with its headlights on, drove ten kilometers away past intervening high ground and out of the line of sight, and presto, the headlights were visible as a Min Min Light. The following morning they took photographs of a distant mountain range distorted by the Fata Morgana, that gradually faded away to its actual position below the horizon as the conditions dispersed.

As long as there have been people around to observe the Min Min Light, there have been people around to build campfires or have lanterns in the window dozens and perhaps hundreds of kilometers away. What about Henry Lamond's account of the light passing him? Well, who knows. Our memories often exaggerate remarkable events. Perhaps his trail curved away and created that perception, perhaps the story was improved in the retelling, there's no way to know.

But whether you choose to regard the Min Min Light as an optical artifact, as the ghost of a cowboy shot in an outback saloon, as a piezoelectric light beacon or as a remote controlled alien probe, there's no doubt that stories like the channel country jack o' lantern make our world ever more intriguing.

REFERENCES & FURTHER READING

Eddingsaas, Nathan C., Suslick, Kenneth S. "Mechanoluminescence: Light from sonication of crystal slurries." *Nature.* 9 Nov. 2006, 444 (7116): 163.

Holden, Alan, Morrison, Phylis. *Crystals and Crystal Growing.* Cambridge: MIT Press, 1982. 225-234.

James, Thomas Horton. *Six Months in South Australia: With some account of Port Philip and Portland Bay in Australia.* London: J. Cross, 1838. 201-202.

Lamond, Henry. "The Min-Min Light." *Walkabout.* 1 Apr. 1937, Volume 3, Number 6: 76-77.

Moravec, Mark. "Strange Illuminations: Min Min Lights Australian Ghost Stories." *Fabula.* 21 May 2003, Volume 44, Number 1: 2-24.

Pettigrew, John D. "The Min Min light and the Fata Morgana: An optical account of a mysterious Australian phenomenon." *Clinical and Experimental Optometry.* 1 Mar. 2003, Volume 86, Number 2: 109-120.

18. The Rendlesham Forest UFO

The Sci-Fi Channel calls it the most comprehensive cover-up in the history of Britain. It's often called the most important UFO incident of the 20th century. Imagine, alien spacecraft drifting through the woods on the perimeter of a US Air Force Base in England, shining their colored lights around in plain view of pursuing military security personnel, for three nights in a row. And how did the United Kingdom and the United States react to this obvious threat to their nuclear arsenals? They didn't. There's no wonder the Rendlesham Forest UFO Incident is the one that UFOlogists consider the most frightening.

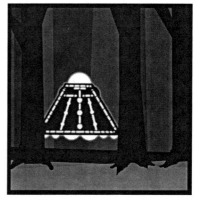

If you watch the Sci-Fi Channel, the History Channel, the Discovery Channel, or any of the other paranormal TV networks, you've probably heard the popular version of events on those three nights. Here are the significant points:

Two old Royal Air Force airfields, RAF Bentwaters and RAF Woodbridge, are situated just two miles apart near the eastern coast of England. Throughout the cold war they were operated by the United States Air Force. On the night of Christmas Day, December 25, 1980, personnel at the base reported bright UFOs streaking through the sky. Later that night, in the wee hours of December 26, security personnel from RAF Woodbridge entered Rendlesham Forest to investigate some strange, pulsating, colored lights moving through the trees, that they thought at first might be a downed aircraft. Local constables were called and also participated in the observation. Base personnel described the craft they pursued as metal and conical, with a bright red light above and a circle of blue lights below, and suspended in a yellow mist. By daylight, they located a clearing where they thought the strangely lit craft had set down, and found three depressions in the ground in a triangular pattern. The constables were called again and photographed and confirmed the landing site.

Two nights later in the wee hours of December 28, they returned to the site, led by Lt. Colonel Charles Halt, second in command at the base. They brought a radiation detector and recorded high levels of radiation at the landing site, again observed the colored, pulsating lights through the trees, and again pursued them through the forest. Other colored lights were seen flying through the sky. Col. Halt recorded the audio of this pursuit on a microcassette. Two weeks later, after debriefing all of his men who participated, he wrote down the specifics of the episode in a signed memo titled "Unexplained Lights", and sent it in to the British Ministry of Defense. Ever since, the airmen involved claim to have been coerced to change their stories and deny that anything happened, and were threatened with comments like "bullets are cheap."

Wow. That story is really something, isn't it? But even more impressive than the story is the documentation, mainly Col. Halt's audio recording and signed memo. You don't rise to be deputy commander of a United States Air Force base with nuclear weapons if you're a nutcase, and when you're accompanied by local police constables and a number of Air Force security personnel who all file written reports, you don't exactly make up ridiculous stories. There's little doubt that Rendlesham Forest probably has the best, most reliable evidence of any popular UFO story.

Ever since I first heard about the Rendlesham Forest incident, I've been as curious as anyone to know what actually happened. So I decided to begin with the null hypothesis — that nothing extraordinary happened — and then examine each piece of evidence that something extraordinary did happen, individually, on its own merit. I wanted to see if we could find a natural explanation for each piece of evidence: You always have to eliminate terrestrial explanations before you can consider the extraterrestrial.

Let's take it chronologically. The first events were the reported UFO sightings at the base on the night of the 25th and the early hours of the 26th. It turns out that people on the base were not the only people to see this. UFO reports flooded in from all over southern England, as it turned out that night was one of the best on record for dramatic meteors. The first were, at 5:20pm and again at 7:20pm over southeastern England. Later at 9pm, the upper stage of a Russian rocket that had launched the Cosmos 749 satellite re-entered and broke up. As reported in the Journal of the British Astronomical Association, 250 people called in and reported a sighting as first six fragments came streaking in, which then broke up into more than 20. Finally, at about 2:45am on the morning of the 26th, a meteor described by witnesses as "bright as the moon" flew overhead with an unusually long duration of 3-4 seconds. The experience of the airmen was described in a letter home written by one of them:

> *At [about 3am], me and five other guys were walking up a dark path about 2 miles from base... Then we saw a bright light go right over us about 50 feet up and just fly over a field. It was silent.*

At the same time on base, a security patrolman was dispatched to check the weapons storage area to see if a "falling star" had hit it. It had not. But it does seem clear that all of the UFO reports from the base are perfectly consistent with known meteor activity on that night. So much for the UFO sightings. Next piece of evidence.

Airmen at the east end of RAF Woodbridge went into the forest to investigate a strange, pulsing, colored light that they suspected might be a downed aircraft. We have the signed statements of the three men who went into the forest, SSgt. Jim Penniston and Airmen Ed Cabansag and John Burroughs, as well as that of their superior, Lt. Fred Buran. At this point it's important to know the geography of the area. Heading east from the east gate of RAF Woodbridge, there is about one mile of forest, followed by an open farmer's field several acres in size. At the far end of that field is a farmhouse. A little more than 5 miles beyond that sits the Orfordness lighthouse, in a direct line of sight.

Although the three men stayed together, their reports are dramatically different. Penniston and Burroughs reported moving lights of different colors, that they felt came from a mechanical object with a red light on top and blue lights below surrounded by a yellow haze. They even drew pictures of it in their reports, but Penniston's illustration of their best view of it shows it partially obscured by trees and well off in the distance to the east. Burroughs' drawing of the object is based on Penniston's description, as Burroughs himself only reported seeing lights. Cabansag, however, reported that the only light they saw after actually leaving the base was the one that all three men eventually identified as a lighthouse or beacon beyond the farmhouse. Cabansag reported that the yellow haze had simply been the glow from the farmhouse lights. Once they reached the field, they turned around and returned to base without further incident.

A further problem with Burroughs' and Penniston's stories is that they have grown substantially over time, particularly Penniston's. In more recent TV interviews, they've both claimed that they saw the craft fly up out of the trees and fly around. Penniston has also unveiled a notebook that he claims he wrote during their forest chase, and which he displayed on a 2003 Sci-Fi Channel documentary. Its times and dates are wrong, and Burroughs has stated that Penniston did not make any notes during the episode and would not have had time to even if he'd wanted. Penniston's story has also expanded to include a 45 minute personal walkaround inspection of the object during which he took a whole roll of photographs (seized by the Air Force, of course), which from the written statements of all three men, is a clear fabrication.

Only Cabansag's version of events, that there was a single pulsing light later determined to be a distant beacon or lighthouse, describes events that all three men agreed on, and is consistent with the statements of others. For example, A1C Chris Arnold, who placed the call to the police and waited at the end of the access road, gave this description in a 1997 interview:

> *There was absolutely nothing in the woods. We could see lights in the distance and it appeared unusual as it was a sweeping light, (we did not know about the lighthouse on the coast at the time). We also saw some strange colored lights in the distance but were unable to determine what they were... Contrary to what some people assert, at the time almost none of us knew there was a lighthouse at Orford Ness. Remember, the vast majority of folks involved were young people, 19, 20, 25 years old. Consequently it wasn't something most of the troops were cognizant of. That's one reason the lights appeared interesting or out of the ordinary to some people.*

Police constables responding to Arnold's call of "unusual lights in the sky" did arrive on the scene while Penniston, Cabansag, and Burroughs were still in the forest. Here is the report they filed:

> *Air traffic control West Drayton checked. No knowledge of aircraft. Reports received of aerial phenomena over southern England during the night. Only lights visible this area was from Orford light house. Search made of area - negative.*

So much for unusual lights or strange flying craft reported by the airmen in the forest on the first night.

Next morning, some of the men found what they believed to be site of where Penniston's craft must have touched down. It was a clearing with three depressions in the ground, possibly made by landing pads. Again the police were called. The police report stated:

> There were three marks in the area which did not follow a set pattern. The impressions made by the marks were of no depth and could have been made by an animal.

Forestry Commission worker Vince Thurkettle, who lived less than a mile away, was also present at the examination of the landing site. Astronomer Ian Ridpath, who has a fantastic web site about the event (and check out this YouTube video of his original BBC report here), interviewed Thurkettle about the impressions and the reported burn marks on the surrounding trees:

> He recognized them as rabbit diggings, several months old and covered with a layer of fallen pine needles... The "burn marks" on the trees were axe cuts in the bark, made by the foresters themselves as a sign that the trees were ready to be felled.

So much for the landing site.

It was two nights later that Col. Halt decided to take the investigation into his own hands (contrary to the popular telling that says there were events on three nights in a row, there are no reported events on the second night). Halt properly armed himself with a Geiger counter and an audio recorder, and took some men to examine the landing site and the strange lights. It's been reported that Halt found radiation levels at the landing site ten times higher than normal background levels:

> Col. Halt: "Up to seven tenths? Or seven units, let's call it, on the point five scale."

He used a standard issue AN-PDR 27 Survey Meter, which detects beta and gamma radiation. The highest level reported by Col. Halt on his audio tape, "seven tenths", corresponded to .07 milliroentgens per hour, just at the lowest reading on the bottom range of the meter, the "point five scale". The UK's National Radiological Protection Board (NRPB) told Ian Ridpath that levels between .05 and .1 mR/h were normal background levels; however, this particular meter was designed to measure much higher levels of radiation and so it was "not credible" to establish a level of only ten times normal background. So much for Col. Halt's radiation.

And then they observed the mysterious colored light flashing through the trees:

> *Col. Halt: "You just saw a light? Where? Slow down. Where?"*
>
> *Unidentified: "Right on this position here. Straight ahead, in between the trees — there it is again. Watch — straight ahead, off my flashlight there, sir. There it is."*
>
> *Col. Halt: "I see it, too. What is it?"*
>
> *Unidentified: "We don't know, sir."*
>
> *Col. Halt: "It's a strange, small red light."*

Every lighthouse has a published interval at which it flashes. This is how sea captains are able to identify which light they're seeing. The Orfordness lighthouse has an interval of 5 seconds. In the above exchange, each time one of the men reported seeing the light, it was exactly five seconds apart.

Although several times during the tape Col. Halt calls the light red, he is contradicted by his men who say it's yellow. In photographs of the 1980 light taken before it was replaced, it did indeed look orange. Even the new light, which is mercury vapor discharge and therefore whiter and bluer than the original incandescent, appears distinctly red in photographs and video when viewed from Rendlesham forest.

Col. Halt, having been in the area longer than most of the young servicemen, did know about the lighthouse; but he didn't think this light could be it because it was coming from the east. Col. Halt believed the lighthouse was to the southeast. This is true from RAF Bentwaters, where Halt was from. But the chase through the forest proceeded due east from RAF Woodbridge — two miles south of Bentwaters — and from there, unknown to Col. Halt, Orfordness lighthouse is indeed due east.

> *Col. Halt: "We've passed the farmer's house and are crossing the next field and now we have multiple sightings of up to five lights with a similar shape and all but they seem to be steady now rather than a pulsating or glow with a red flash."*

Five steady lights glowing red. The Orfordness Transmitting Station is just two miles up the coast from the lighthouse, and features five tall radio towers topped with red lights. Col. Halt's thoroughness was commendable, but even he can be mistaken. Without exception, everything he reported on his audiotape and in his written memo has a perfectly rational and unremarkable explanation.

And with that, we're nearly out of evidence to examine. All that remains is the tale that the men were debriefed and ordered never to mention the event, and warned that "bullets are cheap". Well, as we've seen on television, the men all talk quite freely about it, and even Col. Halt says that to this day nobody has ever debriefed him. So this appears to be just another dramatic invention for television, perhaps from one of the men who have expanded their stories over the years.

When you examine each piece of evidence separately on its own merit, you avoid the trap of pattern matching and finding correlations where none exist. The meteors had nothing to do with the lighthouse or the rabbit diggings, but when you hear all three stories told together, it's easy to conclude (as did the airmen) that the light overhead became an alien spacecraft in the forest. Always remember: Separate pieces of poor evidence don't aggregate together into a single piece of good evidence. You can stack cowpies as high as you want, but they won't turn into a bar of gold.

References & Further Reading

Army, *Radiac Set AN/PDR-27*. Washington, DC: United States Army, 1952. 12.

BBC. "Rendlesham - UFO Hoax." *BBC Inside Out*. BBC, 30 Jun. 2003. Web. 16 Jan. 2010. <http://www.bbc.co.uk/insideout/east/series3/rendlesham_ufos.shtml>

Kelly, Lynne. *The Skeptics Guide to the Paranormal*. New York: Thunder's Mouth Press, 2004. Pages 198-201.

Ridpath, Ian. "The Rendlesham Forest UFO Case." *IanRidpath.com*. Ian Ridpath, 1 Jun. 2009. Web. 16 Jan. 2010. <http://www.ianridpath.com/ufo/rendlesham.htm>

Rutkowski, C. *A World of UFO's*. Toronto: Canada Council for the Arts, 2008. 27-32.

19. Who Is Closed Minded, the Skeptic or the Believer?

The tagline for the Skeptoid podcast, buried somewhere in the RSS feed that nobody ever sees, is "The truth always hurts someone." This axiom is reproven every week when my Google Alerts show me new references on the web to my Skeptoid podcast, usually in somebody's blog, usually lambasting me as a paid stooge for the government or at best, "just another closed-minded debunker". If I don't accept every shred of anecdotal whimsy as absolute proof of the supernatural, I'm "closed minded". If I have not been sufficiently impressed by evidence to move from the null hypothesis, I'm "trying to justify my preconceived notions". If I am moved by the results of well-performed studies that have passed peer review and contribute to a scientific consensus, I've "taken the red pill and gone down the rabbit hole to become a true believer in the lies of Big Science".

Now, take a moment to reflect. Are those not the same things skeptics say about believers? They are, exactly. Skeptics and believers tend to follow the same thought processes, and come to conclusions that validate their own methods and beliefs, and invalidate those of their opponent. More than once, I've had a conversation with a well read, intelligent, articulate true believer, who charged me with the same flaws in my logic that I found glaring in his. I've watched debates between the top names in science and pseudoscience, and seen these conversations deteriorate into little more than "Takes one to know one", "Nuh-uh", and "I know you are, but what am I?"

To be an effective skeptic, it's critical to understand that your opponent is not simply a lunatic. Maybe some are, but the majority are as intelligent and thoughtful as you. Dismissing your opponent as crazy is a weakness in you. When a skeptic talks with a believer, he often finds the believer to be closed minded, in that the believer is not open to any evidence that challenges his belief. The fact is that the believer also finds the skeptic to be closed minded, in that he does not accept the evidence that supports the belief. From the perspective of each, each is right. And that's really important to understand.

Being closed-minded is only one crime of which skeptics and believers accuse each other. Both also accuse each other of being believers, and consider themselves skeptics. For example, a 9/11 Truther honestly believes that what everyone observed on 9/11 is not what happened, and that the consensus of what witnesses, victims, law enforcement, and emergency services experienced on that day is merely a government fabrication. They are skeptical of that fabrication, and thus consider themselves skeptics. They consider those of who accept what was reported on that day to be believers, as if we deliberately close our minds to their hypothesis and instead uncritically accept the "official story".

Similarly, proponents of non-scientific alternatives to healthcare, like reflexology or straight chiropractic or reiki, honestly consider their mistrust of evidence-based medicine to be well-founded skepticism. They consider those of use who "blindly accept what doctors and pharmaceutical companies tell us" to be uncritical believers. From their perspective, that's a reasonable evaluation.

People who watch and believe the ghost hunting shows on television consider what they see to be scientific evidence of ghosts, and that skepticism of the claim that ghosts don't exist is more than justified. They honestly consider those of us who don't buy into those shows to be turning a blind eye to the evidence, and insisting on a stubborn belief that ghosts have never been shown to exist. From their perspective, we are the ones who are closing our minds to the evidence and not being scientific.

I'm not even sure what being "closed minded" is. I guess it means that you won't give a chance to any evidence of any quality. If that's true, then closed minded is probably not a term that genuinely applies to either skeptics or believers. The first step is to be selective about what evidence we turn away at the door, and this is where the real difference is. We all turn away some of the evidence at the door. We have to. It would be impossible to get through your day if you had to devote a full-fledged investigation into every minute suggestion or claim or anecdote that comes along.

And here's an important point. This is a charge that I think we are nearly all guilty of, to some degree: In their process of selecting which evidence to turn away at the door and to which to give further attention, skeptics and believers both tend to select evidence they are likely to be predisposed to accept. Everyone does this. Whether you know anything about the scientific method or not, whether you believe in Bigfoot or not, whether you're trying to justify a preconceived notion or not, and whether you're a skeptic or a believer, we all have our own individual standards by which we select evidence to consider.

This filtering is usually done with a bias, because most people don't happen to be following a formal research protocol every minute of every day. Here's the way this evidence filtering usually plays out in everyday life:

I say: *My dad's diabetes was cured with acupuncture.*

The skeptic says: *We already know there's no plausible connection between the two, I'm not interested in that anecdote.*

The believer says: *I'm very interested in what acupuncture can do, this is one more piece of evidence of its effectiveness.*

I say: *Pfizer just announced its new cancer drug has been approved and found to be safe and effective.*

The skeptic says: *That system's not perfect but it's the best we have, glad to hear we have a new effective option.*

The believer says: *I'm not interested in any claims made by a for-profit pharmaceutical company.*

I say: *Bob saw a UFO and he could see aliens waving through the window.*

The skeptic says: *A fun story, but hardly useful as evidence of anything.*

The believer says: *Taken by itself, this story may not be great evidence, but it adds to the mountain of evidence of alien visitation.*

Even though both of these people follow the same thought process, the skeptic is going to be right more often than the believer. Why? He's no more or less closed minded or biased in the way he shuts out information. The only reason his decisions are usually going to be better is that he has a better general science background. He has a general understanding of what's clinically known about acupuncture, rather than what's in newspaper advertisements. He has a general knowledge of the drug research and approval process. He has a general understanding about quality of evidence and the value of anecdotal evidence.

Here's a real life example. I had a recent email exchange with Stanton Friedman, probably the biggest promoter of alien visitation. He's best known for his promotion of the Roswell mythology, following the *National Enquirer's* 1978 reprinting of the original 31-year-old uncorrected article from the *Roswell Daily Record*. In our emails, Friedman accused me of closing my mind to all the evidence out there. Fair enough; I do filter out evidence that's purely anecdotal, ambiguous, irrelevant, or otherwise useless. So I asked him if he could provide a single piece of useful, unambiguous evidence of alien visitation. He replied in part:

> *I certainly don't have a piece of a saucer. There are about 4,000 physical trace cases from about 90 countries... Having a fingerprint doesn't provide the finger, but proves one was there... We are dealing with intelligently controlled ET spacecraft.*

Friedman is absolutely right that a fingerprint proves a finger was there. A fingerprint is unambiguous and has no other explanation. But what are these 4,000 physical traces? We'd have to look at each one individually and evaluate it; we don't just say "Well 4,000 is a big number, *some* of them must be real." If any are conclusively and unambiguously parts of an alien spaceship, with no other possible explanation, then we'd have something worth looking at. What we don't *ever* do is credit 4,000 pieces of poor evidence in the aggregate as one piece of good evidence. If Friedman followed a responsible research protocol, he would refuse to draw a conclusion from 4,000 useless items of unknown origin. I look at it and see 4,000 arguments that no good evidence has been found. Friedman looks at the same thing and says, and I quote from his email to me, "We are dealing with intelligently controlled ET spacecraft."

So what is the real difference between skeptics and believers? It's disingenuous to claim that either is more closed-minded than the other; everyone sits somewhere along that spectrum, nobody is immune. It's disingenuous to say that either is more of a stubborn believer than the other, or that either tends to support their preconceived notions more than the other. Probably all of us are more guilty of these than we like to admit, especially in less formal environments. The real difference between skeptics and believers is that skeptics have a useful foundation of scientific knowledge and an aptitude for following the scientific method. These tools allow us to distinguish poor quality evidence from good quality evidence. And, importantly, they help restrain us from drawing poorly supported conclusions from the evidence that we do accept, no matter how strongly we want those conclusions to be justified.

So don't focus on buzzword labels like "closed minded" or "true believer". You can be both of those things and still be able to properly analyze evidence and draw a supported conclusion. You can also be guilty of neither fault, and yet be unable to distinguish a well-supported conclusion from mountains of poor evidence. Focus on the method behind the conclusion. Focus on the quality of evidence that supports the conclusion. The ad-hominem attack of "He's closed minded" says nothing at all about the quality of evidence.

References & Further Reading

Achinstein, P. *Scientific evidence: Philosophical theories & applications.* Baltimore: The Johns Hopkins University Press, 2005.

Burton, Robert. *On Being Certain: Believing You Are Right Even When You're Not.* New York: St. Martin's Press, 2008.

Kruglanski, A.W. *The psychology of closed mindedness.* New York: Psychology Press, 2004.

Shamoo, A., Resnik, D. *Responsible Conduct of Research.* Oxford: Oxford University Press, Inc, 2009.

Shermer, M. *How We Believe: Science, Skepticism, and the Search for God.* New York: Henry Holt and Company, LLC, 2000.

20. The Angel of Mons

It was August of 1914, near Mons in Belgium. The German army was making its sweep into France in the opening stages of World War I. Heavily outnumbered units of the British Expeditionary Force came under vastly superior German fire, and their destruction seemed assured. But in perhaps the strangest tale in modern warfare, the British were saved at the last moment by an inexplicable heavenly presence: A brigade of warrior angels appeared and wrought destruction upon the Germans, handing the day and the victory to the British.

At least, that's what you usually hear.

The Angel of Mons was not only a military first, it was also fairly influential in popular culture at the time. Both J.R.R. Tolkien and Mary Norton, author of the *Bedknobs and Broomsticks* trilogy, are said to have been inspired by the story of supernatural soldiers saving the outnumbered good guys from an overwhelming evil enemy force. For decades the story of the Angel of Mons had faded into history, but with the New Age resurgence of angel worship beginning in the 1980's, the story has found its way back into the popular mythology, usually retold without critique. Anyone who's into angel worship can probably rattle off the tale of the Angel of Mons as a great example of guardian angels protecting the good. And it's in the official military records, so you know it has to be true, right?

And that's the perfect place to start our investigation into the Angel of Mons. Contrary to the popular telling, this was not a British victory. In fact those who escaped barely got out with their lives. The Battle of Mons was the first time British and German forces encountered each other in WWI. A large German force was indeed making its sweep into France, and a few small units of the British Expeditionary Force, the first

British soldiers sent to the mainland, happened to be in their way. Outnumbered approximately 4 to 1 at the start of the battle, the British did indeed stop the German advance and inflicted heavy casualties. British infantry were experts on their Lee-Enfield rifles, and many could take a man down at 200 yards at the rate of 15 a minute. The relatively inexperienced and virtually untrained Germans, on the other hand, had no answer for this and believed themselves to be under heavy machine gun fire. The British also used air-bursting shrapnel, which the Germans lacked. After two days of fighting during which the larger German force continuously pushed the British back, the Germans sounded a cease-fire and the British withdrew. British losses in the Battle of Mons were 1,600, and the Germans suffered 5,000 losses. The British had given the Germans a black eye, but the net effect was negligible as it slowed only a small part of the German advance for just two days.

There was no miraculous British victory explainable only by supernatural intervention, and no supernatural intervention appears in any military accounts that I've gone through.

This first clash with the Germans was big news in Britain, as you can well imagine, and new volunteers flooded the recruiting stations when the story broke. In such circumstances it's easy to conceive of infectious patriotism sweeping the land, and the telling of heroic tales, and the trumpeting of news of early victory. Hungry for such stories, a London paper called the *Evening News* engaged Welsh author Arthur Machen, a writer of Gothic horror stories, to publish a tale he called *The Bowmen*. In his story, the besieged British soldiers at the Battle of Mons appealed to St. George for heavenly aid, and who should appear but phantom Medieval longbowmen from the Battle of Agincourt, 500 years past. The heavenly longbowmen decimated the Germans, mysteriously leaving no visible wounds; and carried the day for the British.

> *And as the soldier heard these voices he saw before him, beyond the trench, a long line of shapes, with a shining about*

them. *They were like men who drew the bow, and with another shout their cloud of arrows flew singing and tingling through the air towards the German hosts.*

Pressed by eager editors for more information about this miraculous delivery, Machen was the first to stand up and remind everyone that his was a work of fiction. However, by some accounts, the article had an effect not unlike that of Orson Welles' *War of the Worlds*. It was mistaken for an actual news report by many readers.

The Bowmen was published on September 29, 1914, five weeks after the Battle of Mons. Perhaps the best evidence that the Angel of Mons stories have no factual basis is the absence of any known published accounts referencing supernatural intervention earlier than that date. Except one...

A book published in 1931 by Brigadier-General John Charteris is a collection of letters that he wrote during the war. In one addressed to his wife, dated September 5 (more than three weeks before *The Bowmen)*, he wrote:

Then there is the story of the "Angel of Mons" going strong through the II Corps of how the angel of the Lord on the traditional white horse, and clad all in white with flaming sword, faced the advancing Germans at Mons and forbade their further progress.

The authenticity of Charteris' letter has come under intense scrutiny. For one thing, Charteris served as Chief of British Army Intelligence from 1915 to 1918, and he was involved in numerous schemes to disseminate propaganda. For another, all of Charteris' letters written to his wife during the war were preserved and catalogued by her, and microfilm copies are now kept at the Liddell Hart Centre for Military Archives at King's College, University of London. There is no letter dated September 5, and no letter mentioning any Angel of Mons. Since these archived letters formed the source material for Charteris' 1931 book, researchers like David Clarke have concluded that Charteris falsified this letter after the fact as part of his propaganda duties. He is known to have done this in other cases, notably one where he promoted a false rumor that the Germans were collecting the bodies of their dead and rendering them down in a "cadaver factory" to produce oil and lubricants for their war effort. Charteris' letter has too many strikes against it to be considered reliable evidence that the story of the Angel of Mons existed prior to *The Bowmen*.

For some six months, *The Bowmen* was reprinted not only in newspapers and magazines, but also in spiritual journals; and for that period of time, there was yet no reference in print to angels. Author David Clarke performed an extensive survey of British magazines, newspapers, and journals from the period, searching for the terms "angel of mons" and "angels at Mons". It was not until April 3, 1915, that any mention of angels at Mons first appeared. It was a story from the *Hereford Times* entitled "A Troop of Angels" and gave the report of a young lady named Miss Marrable. She reported stories she'd heard from soldiers who were at the battle:

> *Last Sunday I met Miss Marrable [who] knew the officers, both of whom had themselves seen the Angels, who saved our left wing from the Germans when they came right upon them during our retreat from Mons... One of Miss Marrable's friends, who was not a religious man, told her he saw a troop of Angels between us and the enemy, and has been a changed man ever since. The other man she met in London last week [said that] while he and his company were retreating, they heard the German cavalry tearing after them... They turned around and faced the enemy, expecting instant death; when, to their wonder, they saw between them and the enemy a whole troop of Angels. And the horses of the Germans turned around, terrified out of their senses, and stampeded.*

"A Troop of Angels" was then broadly reprinted, most influentially in May 1915 in *The All Saints (Clifton) Parish Magazine*. But when Miss Marrable was sought out for more information, she said she'd been misquoted. None of the soldiers in her story were named, but some soldiers began coming forward saying things like they knew someone who met someone who heard the story from their very reliable friend. Author Harold

Begbie published *On the Side of the Angels*, in which he charged Arthur Machen with exploiting the true story of angels for his own financial gain. Machen challenged Begbie to then produce these witnesses, and Begbie countered that a government coverup had silenced them. Anyway, suffice it to say that beyond these hearsay accounts in newspapers, no reliable evidence or witnesses were ever produced that could corroborate stories of anything unusual happening at Mons.

And that's just on the British side. I asked two friends in Germany to research whatever records they could find, and they came up completely empty handed, beyond the information already discussed. If the story was exploited by the British army for propaganda purposes, the same certainly wasn't true on the German side of the lines. We couldn't find any record of Germans reporting being chased away by angels or shot by Medieval archers, and you can't plausibly credit a British government coverup for that.

We can't say for certain that no angels turned the tide of battle at Mons, but we do have two items that make it highly improbable: First, that no reliable records exist of it ever having happened; and second, that its genesis as a story is well documented as fiction and as derivative reporting based on that fiction. So enjoy your *Bedknobs and Broomsticks*, but don't spend too much time looking for a historical basis.

References & Further Reading

Begbie, H. *On the side of the Angels.* London: Hodder and Stoughton, 1915. 7.

Clarke, D. *The Angel of Mons: Phantom Soldiers and Ghostly Guardians.* West Sussex: John Wiley and Sons Ltd., 2004.

Dupuy, R., Dupuy, T. *The Harper Encyclopedia of Military History, Fourth Edition.* New York: HarperCollins Publishers, 1993. 1025-1026.

Machen, A. "Short Stories reflecting the times." *Aftermath.* Aftermath, 1 Jan. 2009. Web. 19 Jan. 2009. <http://www.aftermathww1.com/bowmen.asp>

Perris, G. "Constructs battle along the Meuse." *Winnipeg Free Press.* 27 Aug. 1914, Volume 41, No. 46: 1-2.

21. Is He Real, or Is He Fictional?

Let's take a break today from serious investigation and take a walk on the lighter side. There are many well-known characters in popular culture, many who became famous in works of fiction. But some of them are based on real people who actually lived. Most people will probably know most of these, but I bet you nobody will know all of them. Let's get started with some really easy ones, from ancient history; starting with:

Norse hero **Beowulf:** Fictional. This hero of the Old English poem of the same name is said to have lived some 1,500 years ago, 500 years before the great poem was written about his battles with the monster Grendel and other creatures. There is no historical reference to any such person having actually lived, outside of literature. And thankfully, no historical reference to any of the monsters either.

Ancient Greek hero **Ulysses** aka **Odysseus:** Probably fictional. Although history can't tell us for certain whether there was an actual Greek king of Ithaca named Odysseus, we also don't have reason to believe there wasn't. Certainly the tales told about him in Homer's *The Iliad* and *The Odyssey* were purely fiction, but they were also full of real people, places, and events. Surprisingly, it's actually more likely that Odysseus himself was a real person than his storyteller, Homer, who is considered by most scholars to have been a fictitious name attached to the works of multiple poets.

Persian adventurer **Sinbad the Sailor:** Fictional. The only seven voyages this seagoing adventurer ever took were in Western translations of the book *One Thousand and One Nights*. It's not known who originally wrote the tale of this ancient Persian adventurer, but along with his fellow popular characters Aladdin and Ali Baba, Sinbad never actually appeared in Arabic versions of the *Nights*.

You probably got all of those. So let's move on to American history and see if you can keep your streak going, beginning with:

American icon **Uncle Sam**: Real. Sam Wilson owned a meat-packing company that sold barrels of beef to U.S. soldiers during the War of 1812, stamped "US" which the soldiers joked must have stood for Uncle Sam. His reputation was that of a man of great character and honesty, and 150 years later in 1961, an act of Congress officially saluted "Uncle Sam Wilson" as the progenitor of America's national symbol.

Tree-planting folk character **Johnny Appleseed**: Real. John Chapman earned the nickname Johnny Appleseed by planting nurseries to grow apple trees, beginning on his own land grant that he received as a revolutionary war soldier fighting under George Washington. A man of great piety and faith, he lived an almost hermit-like lifestyle of service. He founded nurseries throughout northern Ohio and encouraged his managers to sell or give away the trees as cheaply as possible.

Railroad hammer man **John Henry**: Fictional. This railroad spike-driving hero of folklore is said to have worked himself to death winning a contest against the new steam hammer. Although such a contest may have actually happened in the 1870's or 1880's, and although there were hammer men named John Henry, attempts to reconcile the name with the time, place, and event have been post-hoc efforts.

Frontier adventurer **Daniel Boone**: Real. Often confused with Davy Crockett, the Congressman who died at The Alamo, Daniel Boone lived 50 years earlier. He was a Revolutionary War soldier who went on to blaze a trail for 200,000 pioneers into Kentucky. Despite fighting many Indian battles, Boone actually lived with the Shawnee Indians in Kentucky for some time.

Indian maiden **Pocahontas**: Real. The daughter of the Powhatan chief did indeed marry an Englishman and travel to England in 1616 where she promptly died of pneumonia, but she didn't marry Captain John Smith as many think. She married the tobacco pioneer John Rolfe. Captain Smith did tell how Pocahontas successfully begged her father to spare his life, but although heavily romanticized in fiction, the only evidence that such an event ever took place was Smith's own dubious account.

Hero of folk song **Tom Dooley**: Real. The popular folk song is based on Tom Dula, a Confederate soldier who returned home after the war and was famously convicted and hanged for the murder of his fiance Laura Foster. For over a century, historians have fruitlessly debated whether he was guilty, or whether the true killer was Laura's sister Ann, Tom's first love, and who is said to have confessed to the murder on her deathbed.

Railroad engineer **Casey Jones**: Real. John Jones, a train engineer from Cayce, Kentucky, died in 1900 trying to stop his train before a collision. While others jumped, he stayed at the brake and was the only person killed. Casey was known for such heroics throughout his career, including one time when he swung out onto the cowcatcher and snatched a frightened little girl from the tracks.

Lumberjack **Paul Bunyan**: Fictional. Tall tales from the logging industry were first retold in print in 1910 by James McGillivray and other writers. What scholarly doubt exists is not so much whether there was a real logger named Paul Bunyan (there wasn't), but whether even the stories themselves ever actually existed among lumbermen at all; or were simply made up by the authors.

Dueling families **The Hatfields and the McCoys**: Real. Sometimes confused with the fictional feud between the Grangerford and Shepherdson clans from *The Adventures of Huckleberry Finn*, the Hatfields and the McCoys were real. In the 1880's, the Hatfields lived on the West Virginia side of the Tug Fork river, and the McCoys lived on the Kentucky side. The McCoys got the worst of it, losing nine killed; until the law put a stop to it by arresting eight Hatfields, hanging one and imprisoning the rest for life.

Now let's cross the pond and take a look into European personalities. European history is a lot older than that of the Americas, so there's been much more time for tales to grow taller and annals to become obfuscated in time and retelling.

Archer and outlaw **Robin Hood**: Fictional. For hundreds of years, minstrels have been singing ballads of the legendary outlaw of Sherwood Forest in Nottinghamshire with almost supernatural archery skills. Much scholarly work has tried to prove that he was real, but there is simply no good evidence. There was (and still is) an Earl of Huntingdon, Robin Hood's legitimate title; but the lineage of the title is well documented and includes no outlaws. One complication is that Robert or Robin Hood were very common names, and certainly some of them were on the wrong side of the law.

Grail-questing sovereign **King Arthur**: Fictional. Historians have consistently failed to find any Arthur Pendragon wielding a sword named Excalibur among the known principal leaders in Britain during the 6th century, although there are good candidates for much of the Arthurian legend. One is a 2nd century Roman officer in Britain, Lucius Artorius Castus, who commanded armored soldiers who fought with swords and lances on horseback, beneath a dragon-head (or Penn-dragon) banner.

Marooned sailor **Robinson Crusoe**: Fictional. The star of Daniel DeFoe's castaway novel is completely made up. People keep trying to point out actual castaways as "the inspiration" for Robinson Crusoe, but it's not like it's so abstract a concept that DeFoe needed a true story to copy from. Most plausibly, the father of DeFoe's publisher had previously published a book by one Henry Pitman who escaped a penal colony only to be marooned on an island. DeFoe and Pitman may have actually met, giving DeFoe just such a true story first hand.

Famous prisoner **The Man in the Iron Mask:** Real. This state prisoner of Louis XIV lived out his life in the Bastille and other prisons, though he was well treated and lived in great comfort. Letters indicate that he probably only wore a mask when he was transported, and it was probably made of black velvet, not iron. There are dozens of theories of who he might have been, but the one most historians agree is false is the best known, that he was Louis XIV's twin brother.

Keeper of the ocean's dead **Davy Jones:** Fictional. For centuries, sailors have spoken of "going to Davy Jones' locker" at the bottom of the sea. He's the mariner's version of the devil. History is full of sea captains, pirates, and tavern keepers named Davy Jones, but none ever achieved any particular note; and despite a fair number of reasonable sounding theories, there is no clear Davy Jones in history that would be a good match for this legend.

Archer **William Tell:** Fictional. Although most Swiss believe their national hero actually did take his famous shot in 1307, historians have simply found too many other versions of the exact same tale in other cultures from other centuries to give the William Tell version any special credence. In these tales, the archer disrespects an official's hat by shooting it, and as punishment, is made to shoot an apple from his son's head. He draws two arrows, intending to kill the official if he misses. Whether or not all these stories stem from an actual event has been long lost to history.

Rat exterminator the **Pied Piper:** Fictional. According to an ancient stained glass window in a church in Hamelin, Germany, the Pied Piper played his fife and lured all the rats out of town and drowned them in the river. But the town refused to pay his bill, and so he returned and played his fife again, this time luring all the children from the town. The town chronicles say in 1284 that "it is ten years since our children left," but no record survives that says what might have happened, and for which the Pied Piper story is a presumed allegory. Theories include disease, an accident, or emigration.

Detective **Sherlock Holmes:** Fictional. Although purely the literary invention of author Sir Arthur Conan Doyle, Sherlock Holmes is often believed to have been a real private eye. He's been incorporated into so many other authors' works as if he were a real person that the confusion is understandable. Mark Twain employed Sherlock Holmes as freely as he did actual characters from history, and a number of Sherlock Holmes biographies and family histories have been written.

It makes you wonder if many years after you die, people will wonder if you ever existed or were just a story. In a thousand years who's going to know how many Hollywood movie characters were based on real life people? *Spinal Tap* might end up being remembered as one of history's great rock bands, and children might sing nursery rhymes about John F. Kennedy. It all goes to show yet again that no matter how sure you are about something, it always pays to be skeptical.

References & Further Reading

Morgan, R. *Boone: A Biography.* Chapel Hill: Algonquin Books of Chapel Hill, 2007.

Nelson, S.R. *Steel Drivin' Man: John Henry, the Untold Story of an American Legend.* New York: Oxford University Press, Inc, 2006.

Price, R. "The New England Origins of "Johnny Appleseed"." *The New England Quarterly.* 1 Sep. 1939, Volume 12, Number 3: 454-469.

Tilton, R.S. *Pocahontas: The Evolution of an American Narrative.* New York: Cambridge University Press, 1994.

Ward, A. W., Waller, A. R., Editors. *The Cambridge History of English Literature. Volume I. From the Beginnings to the Cycles of Romance.* New York: G. P. Putnam's Sons, 1908. 27.

West, J.F. *The Ballad of Tom Dula.* Boone: Parkway Publishers, Inc., 2002.

22. The Case of the Strange Skulls

Is it an extraterrestrial alien? Is it an unknown subspecies of human? Is it a relic from a mysterious race of subterranean or Atlantean beings? Or, could there be some other, more natural explanation for the many strange misshapen skulls found throughout the world? Let's look at some of the most popular:

The Starchild Skull

Perhaps the most famous of the "strange skulls" is the so-called "Starchild Skull", the skull of a five-ear-old child found alongside a normal adult in a cave in Mexico, carbon dated to about 900 years old. Although only the top half of the skull remains, its severe deformities are clear. The skull is exceptionally broad and bulbous, not unlike modern concept sketches of "gray aliens". For this reason, the skull's current owner, Ray Young and author Lloyd Pye promote the skull as alien, or possibly an alien-human hybrid.

Neurologist Steven Novella has pointed out the probable consensus among medical professionals who have seen pictures of the Starchild, that it is likely an unfortunate case of hydrocephaly. Hydrocephaly is a condition where too much fluid accumulates inside the skull of a young child, causing the skull (which is still pliable at that age) to expand. This condition is well established to produce precisely the type of deformity seen in the Starchild skull. Victims of hydrocephaly rarely live beyond the Starchild's young age. Of course it's not possible to make a conclusive diagnosis without a direct examination. Young and Pye list a handful of specialists who they say have viewed the skull or its X-rays, but none of them (so far as I could tell) concur with the speculation that the skull is alien.

The primary strike against them is the 2003 DNA analysis. The child had normal human DNA and was male, having both X and Y chromosomes, proving that it had a human female mother and a human male father. Young and Pye state that this is consistent with the alien-human hybrid theory, which it may be, who knows; no known alien DNA exists to use as a reference. But it is also consistent with a human child with a well documented and thoroughly understood illness.

The Peruvian Coneheads

Author Robert Connolly has collected and photographed a number of skulls from ancient Peru, skulls that are surprisingly elongated. In Peru, a practice called "skull binding" involved wrapping fabric or leather straps about a child's head, molding it as it grew into this strange oval shape. There's nothing mysterious or unknown about this; we have plenty of historical and archaeological data about this practice. But Connolly disagrees, stating that this explanation "has been rejected". Well, it has, by him; but not by the anthropologists who make Peru their business. The elongated skulls of Peru are certainly interesting, but their origin is well understood and no mystery exists outside of the delusions of those who insist on alien or supernatural explanations for just about anything.

The American Giants

The next skulls I wanted to discuss were the strange giant skulls from Minnesota with double rows of teeth. But I quickly ran into a problem: For virtually the entire 19th century, people were reporting discoveries of skeletons with double teeth all over the United States. Double teeth sounds interesting. I found a number of repeated references that the Talmud states that some Biblical giants had double rows of teeth, but in an online searchable Talmud I found no such reference. It's also said that the Fomorians, an early Irish tribe, are known to have had double teeth. In fact, the Fomorians were a fictional race of ogres, and even so I found no mention of double teeth outside of web pages promoting strange skulls. Here are several of the double teeth stories:

- In Clearwater, Minnesota in 1888, seven skeletons were found buried in an upside-down sitting position facing a lake, all with sloping foreheads and double rows of teeth on the top and bottom.
- In Jefferson County, New York in 1878, a skeleton of "great stature" was found buried under the roots of a large maple, with entire rows of double teeth; found among the skeletons of hundreds of men, who apparently fell defending the ditch they were in.
- Proctorville, Ohio, 1892. A "very large" skull was unearthed with double teeth.
- Medina County, Ohio, 1881. Nine skeletons found while digging the cellar of a house, all with double teeth. The skulls were so large one man was able to put one on like a hat, and it rested upon his shoulders.
- Virginia, 1845, a human jawbone with transverse teeth, so large it could wrap around a man's face.
- At an Indian burial ground excavated in Michigan in 1890, three of the skulls recovered had three holes in the crown, and of these, two had double teeth in front.
- In 1829, in Chesterton, Ohio, a mound was excavated in which was found a jawbone with "more teeth" than modern humans, and which was so large that it could fit over the face of a man known for his large jaw.
- Noble County, Ohio, 1872: Three skeletons, at least eight feet in height, recovered from a mound. The skulls all had double teeth, and "upon exposure to the atmosphere, the skeletons soon crumbled back to Earth."

- Rancho Lompoc, California, 1833: Soldiers digging a pit discovered a twelve foot tall skeleton, its skull featuring double rows of teeth on the upper and lower jaws.

- A frequent visitor to Santa Rosa Island off California in 1860 often discovered numerous Indian skeletons in caves, many of which had double teeth all around.

- Mason County, Virginia, 1821: Seven skulls found, so large they could easily fit over a man's head; with double rows of teeth on the upper jaw, and but two solitary teeth on the lower jaw.

Besides the extra teeth, all of these stories have one very important fact in common: A complete lack of evidence. No photographs, and certainly no skulls in museums or private collections, at least not that I could find any record of. If you found an extraordinary skull, wouldn't you keep it? Wouldn't you show it to some professors at the local college? At a minimum, wouldn't you sell such an amazing find to someone who would display it or preserve it? But in every one of these cases, no record of the skull exists at all. And although it seems every other thrust of a shovel used to turn one of these up, apparently not a single one has been found since 1892, despite far more extensive construction and excavation since then. Even if these were hoaxes, someone would have photographed it or sketched it or preserved it. For the double-toothed skulls, I've found no reason to move from the null hypothesis that no such skull has ever actually been found. I'd love to be proven wrong.

The stories are symptomatic of a late 19th century fad. Strange skeletons and petrified people were quite popular in the United States in the 19th century, and it wasn't just limited to skulls with double teeth. It seemed that just about everyone with a circus tent or a traveling show had some enigmatic human remains, which always conveniently managed to be lost or destroyed before they could be properly scrutinized. Mark Twain lampooned these stories on a number of occasions. He once wrote:

> *In the fall of 1862, in Nevada and California, the people got to running wild about extraordinary petrifactions and other natural marvels. One could scarcely pick up a paper without finding in it one or two glorified discoveries of this kind. The mania was becoming a little ridiculous. I was a brand-new local editor in Virginia City, and I felt called upon to destroy this growing evil... I chose to kill the petrifaction mania with a delicate, a very delicate satire. But maybe it was altogether too delicate, for nobody ever perceived the satire part of it at all. I put my scheme in the shape of the discovery of a remarkably petrified man.*

He then went on to describe the discovery of a petrified Indian making an obscene gesture. He also wrote a tale called *A Ghost Story* in which he is visited by the ghost of the Cardiff Giant, a 10-foot petrified man discovered in 1869 and proven to be a hoax. Twain explains to the ghost that he's been haunting P.T. Barnum's plaster copy of the Cardiff Giant instead of the original.

It's wise to understand the carnival-like atmosphere in which many of the 19th century freakshow skeletons are said to have been discovered, and the convenient lack of evidence. A few of the many stories that fit this model are:

- Skeletons were exhumed from a burial mound in Bradford County, Pennsylvania, in the 1880's. They were seven feet tall, but most extraordinarily, they had horns. Although modern replicas have been created and photographed based on the descriptions, the original horned skulls were sent to the American Investigating Museum in Philadelphia where they vanished before they could be documented. (Incidentally, the entire Internet has not a single reference to any institution called the "American Investigating Museum" outside of this one story.)
- Near Coshocton, Ohio in 1837, a number of "pygmy" skeletons are said to have been found. Various reports give the number as being "several" on the low end, to "very numerous" and "tenants of a large city" on the high end. Remains of wood found near some

of them suggested that they may have originally been buried in coffins. The skeletons were three to four feet tall, and no objects were found that may have helped identify them. Unfortunately the skeletons were "reduced to chalky ashes" and could not be preserved or documented.

- A mound excavated in Ohio in 1891 revealed the skeleton known as "Copper Man", so named because of the copper jewelry and other artifacts buried with him, and said to be "enormous". Strange skull aficionados love Copper Man in part because he was actually written up in the December 17, 1891 edition of the journal Nature, issue 1155. Debunk that! In fact, the Nature article was actually a review of a book about gigantism and acromegaly that referenced the alleged discovery, and was not a peer-reviewed report of the find itself.

- A Mr. Robinson excavated a burial mound on his property in Brewersville, Indiana in either 1879 or 1891 and found a skeleton said to be nine and a half feet tall. Although Mr. Robinson declared that it had been examined by scientists from Indiana and from New York, the bones were suddenly washed away in a 1937 flood before they could be documented.

- An 1876 report mentioned in the 1978 book *A Handbook of Puzzling Artifacts* alleges that James Brown was plowing his field in Coffee County, Tennessee when he began turning up numerous skeletons of three foot tall pygmies. It was estimated that 75,000 to 100,000 skeletons were buried in his six acres. Sadly, not a single one of them was preserved or documented.

- Railroad workers in Hardin County, Ohio removed a 150-year-old oak stump in 1856, uncovering the bones of a man so large his vertebrae were said to be the size of those of a horse. Although the supervisor had noted they were quarrying from a burial mound and thus took appropriate caution and notes, it did not seem to occur to him to remove or preserve the extraordinary find.

Mound building was indeed common among early pre-Columbian cultures in North America, going back as far as 3,000 BCE. Thousands of such mounds have been discovered throughout the Ohio and Mississippi valleys. Metal artifacts and human remains have been discovered, but so far nothing that has surprised archaeologists: No double teeth, no pygmies, no giants. However, cultures from the early American Hopewell tradition did have a couple of practices that may account for many of these stories. They would often disarticulate the bones of the dead for burial, laying them out separated from one another, such that an inexperienced discoverer may mistake their arrangement for the remains of a person seven, eight, or even nine feet tall. Jewelry was also made from bones, notably jawbones; and this involved the boring of extra holes to accommodate leather strings. These extra holes bored in jawbones could easily have been mistaken for sockets for extra rows of teeth.

In the midst of such intriguing mounds, it's easy to see how farmers and fortune seekers might try to generate a little sensationalism with a wild story or two. The anecdotal evidence in the form of stories has great value to investigators looking to support the claims, but so far, these stories have led to a grand total of strange skulls of exactly zero. Always remember: No matter how much poor-quality evidence you have, it does not aggregate into a single piece of good-quality evidence. You can stack cowpies as high as you want; they won't turn into a bar of gold. When you hear lots of stories that are supported by dubious or nonexistent evidence, no matter how many there are, you have good reason to be skeptical.

REFERENCES & FURTHER READING

Castriota-Scanderbeg, A., Dallapiccola, B. *Abnormal Skeletal Phenotypes*. Berlin, Germany: Springer, 2005. 3-100, 501-931.

Gerszten, P.C. "An Investigation into the Practice of Cranial Deformation Among the Pre-Columbian Peoples of Northern Chile." *International Journal of Osteoarchaeology.* 27 May 2005, Volume 3, Issue 2: 87-98.

Hippocrates; translated by Adams, F. *On Airs, Waters, and Places.* Gloucester, UK: Dodo Press, 2009. 20.

Kirks, D. (Editor). *Practical pediatric imaging: Diagnostic radiology of infants and children (3rd Edition).* Philadelphia: Lippincott-Raven, 1998. 80-200.

Novella, S. "The Starchild Project." *The New England Skeptical Society.* The New England Skeptical Society, 1 Feb. 2006. Web. 14 Jan. 2010. <http://www.theness.com/the-starchild-project/>

Smith, B. et. al. "An anatomical study of a duplication 6p based on two sibs." *American Journal of Medical Genetics.* 3 Jun. 2005, Volume 20, Issue 4: 649-663.

23. Kangen Water: Change Your Water, Change Your Life

Today we're going to take a scientific look at one of the latest multilevel marketing fads: healing water machines, devices costing thousands of dollars claiming to ionize or alkalize your tap water, and claiming a dazzling range of health and medical benefits. Sold under such names as Kangen, Jupiter Science, KYK, and literally hundreds of others, these machines do either nothing or almost nothing (beyond basic water filtration), and none of what they may actually do has any plausible beneficial purpose. They are built around the central notion that regular water is so harmful to the body that their price tags, as much as $6,000, are actually justified. They are essentially water filters with some additional electronics to perform electrolysis. They are sold with volumes of technical sounding babble that may impress a non-scientific layperson, but to any chemist or medical doctor, they are laughably meaningless (and in many cases, outright wrong).

Here's a really quick coverage of the basics of the real science. The pH scale goes from about 0 to 14. 7 is neutral pH. Lower numbers are acidic, and higher numbers are alkaline. All aqueous solutions contain some dissociated water molecules in the form of positive hydrogen ions (H^+) and negative hydroxide ions (OH^-). When there are more hydroxide ions, it's because the solution contains positively charged metal ions like sodium, calcium or magnesium for those hydroxide ions to bind to, thus making the solution alkaline. Conversely, when there are more positive hydrogen ions, there needs to be some other negatively charged ions, usually bicarbonate (HCO_3^-) and the solution is thus acidic. Pure water has neither such chemicals in it, and so it has neutral pH. To electrolyze or ionize water, you *must* add some chemicals of one type or the other.

Make no mistake about it: Ionizing and alkalizing water machines are a textbook example of inventing an imaginary problem that needs to be solved with expensive pseudoscientific hardware. It should come as no surprise that the most expensive of these machines are usually sold through multilevel marketing: A one-two punch that first takes advantage of a layperson's lack of scientific expertise to interest them in the product, and then takes advantage of their lack of business or mathematical expertise to convince them that they're virtually guaranteed to become a millionaire through a pyramid model.

The company making the most noise right now is Kangen. They use the slogan "Change your water - change your life." Google that phrase; 49 million results currently. It's a brilliant slogan; everyone would like to change their life for the better, wouldn't it be great if all it took was changing your water? I glance over some of these URLs: MyMiracleWater.com, VeryHealthyWater.com, WaterMiracles.com, AlkalineWaterMiracle.info — people selling easy answers to imaginary problems.

Let's look at the claims these sellers make. The successful MLM companies generally dodge government regulators by making no illegal claims themselves; instead, they allow those claims to be made by their independent distributors: First charging them big dollars for the privilege, and then burdening them with the risk of needing to make untrue health claims in order to recoup their foolish investment. So I've looked over a lot of these independent web sites and come up with what they generally say are the reasons you need to buy their supposedly special water.

Ionized water molecules form into hexagonal rings, which allow the water to be better absorbed by your body.

Water molecules in liquid water move about freely, there is no way that a hexagonal arrangement could be formed or made stable. Stephen Lower is one of many chemists who have debunked this completely made-up and chemically implausible claim. If you're interested in the details, read his excellent web page "Water Cluster Quackery". Hexagonal arrangements of liquid molecules are not a characteristic of ionization or of alkalinity. Such hexagonal arrangements in water have never been observed or plausibly theorized, and thus there is no way that it could have ever been established that such water is better absorbed by your body — since it doesn't exist. The human body has never had a problem being hydrated by water, so this particular claim is a perfect example of a pseudoscientific solution to an imaginary problem.

Kangen water is ionized, which makes it alkaline.

Pure water actually cannot be electrolyzed and dissociated into ions to any appreciable degree, it's not electrically conductive enough. You need to have a significant amount of minerals and impurities in order for it to be electrolyzed, which is why Kangen and its competitors also take your money for packets of mineral salt additives that you need to add to your water to make your machine do anything. Do this, and your water will become chemically alkaline with a cargo of dissolved metallic ions in solution. Basically, your $6,000 Kangen machine, when used with the provided chemicals, is a way to accomplish the same thing as making a weak Clorox bleach solution. To chemists, the term "ionized water" is meaningless.

Alkaline water promotes healthy weight loss, and boosts the immune system.

Two scientific-sounding medical claims, both too vague to be testable. "Immune system boosting" is medically meaningless. Basically, you can't be healthier than healthy; and a healthy immune system is a delicate balance between attacking foreign bodies and attacking your own healthy tissue. "Boosting" it, if such were possible, would cause your own healthy tissue to be attacked. This is called an autoimmune disease, such as lupus. It's not something you *want*. Alkaline water has never been shown to have any such effect.

Alkaline water is an antioxidant that neutralizes free radicals and slows the aging process.

Although oxidation does contribute to some age-related diseases, consuming antioxidants does not affect normal aging. Even if they did, you wouldn't get them from alkalized water: When water is alkalized, it contains hypochlorites, which are oxidizing agents. Basically, the opposite of what is claimed.

Drinking alkaline water reduces the acidity in your body and restores it to a healthy alkaline state. It is well known in the medical community that an overly acidic body is the root of many common diseases, such as obesity, osteoporosis, diabetes, high blood-pressure and more.

This is absolutely false. Your body's acidity is not, in any way, affected by the pH of what you eat or drink. Eating alkaline food stimulates production of acidic digestive enzymes, and eating acidic foods causes the stomach to produce fewer acids. Your body's primary mechanism for the control of pH is the exhalation of carbon dioxide, which governs the amount of carbonic acid in the blood. Nor has there ever been any plausible research that shows any connection between these diseases and body acidity; this also appears to be completely made up. This is a classic case of using simplistic terminology to sell a product to the scientifically illiterate.

Alkaline water detoxifies and cleanses your colon. Without it, mucoid plaque clogs your bowels and contributes to many diseases.

The dreaded mucoid plaque! Mucoid plaque is an invention by the purveyors of colon cleansing products, it has never actually been observed in medical science. Since it doesn't exist, it's impossible to say whether it would be affected by an alkaline diet. Digestive enzymes neutralize the pH of whatever you eat by the time it gets to your bowels anyway, so it's hard to imagine what science might possibly support a claim such as this.

Kangen water is an anti-bacterial cleanser. Kills 99% of bacteria on contact. Spray it on your throat to prevent a cold.

Fascinating. They also promote Kangen water to aquarium owners because of its "amazing" power to *support* bacteria! The fact is that some bacteria are affected by pH, and some are not. Most thrive in a particular range, but relatively few bacteria are affected by the small 1 or 2 point difference between tap water and water that has been treated with Kangen mineral salt additives. It could be argued that sellers are simply saying whatever they think their target market wants to hear.

Acidic water, like that from your tap, is harmful.

The most common source of acidic water is the cleanest and most natural of all: normal rainwater, with a pH of about 5.6. Most tap water is within a point of 7, which is neutral, so your tap water is probably already more alkaline than clean rainwater. Are you still convinced that this is so dangerous that you need to drop two to six thousand dollars on a machine that any chemist or dietitian will tell you has no credible benefit?

There is one possible use for water if it could be made heavily alkaline, and that's to treat heartburn in the esophagus. But it wouldn't be anywhere near as effective as, for example, a single Tums tablet. However, water so treated would have to be so laden with salts that it would be virtually undrinkable.

Please, everyone: Before you invest money in a Kangen machine or any similar competitive machine, or in becoming a distributor for them, do two things. First, ask a chemist to review their scientific claims; and second, ask a doctor about the medical claims. Maybe you'll find that I'm wrong and the multilevel marketing people have discovered whole new branches of chemistry and medicine heretofore unknown to science. Or maybe you'll find that they're simply another spin-the-wheel-and-invent-a-new-age-pseudoscience trying to separate you from your money with fantastic technobabble and glamorous personal testimonials, and just maybe you'll save those thousands of dollars.

References & Further Reading

Bender, D.A. "The Crystal Truth about Ionized Water." *Health Watch.* 1 Apr. 2005, Newsletter 57: 8.

Hanaoka, K. "Antioxidant effects of reduced water produced by electrolysis of sodium chloride solutions." *Journal of Applied Electrochemistry.* 21 Aug. 2001, 31: 1307–1313.

Hiraoka, A., Takemoto, M., Suzuki, S., Shinohara, A., Chiba, M., Shirao, M., Yoshimura, Y. "Studies on the Properties and Real Existence of Aqueous Solution Systems that are Assumed to have Antioxidant Activities by the Action of "Active Hydrogen"." *Journal of Health Science.* 1 Jan. 2004, Volume 50, Number 5: 456-465.

Lower, Stephen. "'Ionized' and Alkaline Water." *Water Pseudoscience and Quackery.* AquaScams, 11 May 2009. Web. 14 Jan. 2010. <http://www.chem1.com/CQ/ionbunk.html>

Melton, Lisa. "The antioxidant myth: a medical fairy tale." *New Scientist.* 5 Aug. 2006, Issue Number 2563: 40-43.

Uthman, E. "Mucoid Plaque." *Quackwatch.* Stephen Barrett, MD, 7 Jan. 1998. Web. 3 Feb. 2009. <http://www.quackwatch.org/04ConsumerEducation/QA/mucoidplaque.html>

24. The Bosnian Pyramids

Deep inside Bosnia, in a small village called Visoko, just 15 miles northwest of Sarajevo, stands the world's largest and oldest ancient pyramid.

Or so it would appear, anyway. Ancient pyramids are not completely unknown in Europe. The remains of a small pyramid stand in France, two are found in Greece, and one in Italy. These aren't much larger than a house and certainly nothing like the massive structures in Egypt and Latin America. Thus, claims that a giant pyramid to dwarf them all stands in broad daylight in Bosnia are sure to raise eyebrows.

The pyramid sprang onto the world scene in October of 2005, when Semir Osmanagić, a Bosnian who lives in the United States and owns a successful metalworking shop, announced his discovery. The hill, named Visočica, is at the south end of Visoko and although it's completely covered in trees and other greenery, two sides of it are indeed remarkably pyramid-like. If you look at it in 3D you can see that its two clear sides are surprisingly flat, and with a well defined edge between them. These two sides appear to have a nearly square base. The sides face nearly exactly north and east. A 1973 photo shows a great view of Visočica, and though it's a little more squat than what we've seen in Egypt, you clearly don't have to be a crazy person to see a pyramid in its shape.

Just don't look too carefully at the other two sides of the hill, which don't bear any resemblance to a pyramid; they just blend into the natural contours of the other hills surrounding Visočica. This explains why all of the photos you see of Visočica are taken from the same angle — the only one that looks good.

Osmanagić does have his detractors, and it's quite bit more than archaeologists snickering at his claims behind his back. Apparently Visočica happens to also be a real archaeological site. A Medieval fort called Visoki has been excavated on the summit of the hill and declared a National Monument. It's built on top of Roman ruins, which were in turn built on top of ruins from an even older tribe called the Illyirians. Osmanagić's digging amid these ruins, looking for pyramids, has caused outrage in the archaeological community, and many have been lobbying to have his digging permit revoked. In fact, the European Association of Archaeologists published an official statement signed by the presidents and/or directors of the official archaeological organizations of seven European nations that said:

> *We, the undersigned professional archaeologists from all parts of Europe, wish to protest strongly at the continuing support by the Bosnian authorities for the so-called "pyramid" project being conducted on hills at and near Visoko. This scheme is a cruel hoax on an unsuspecting public and has no place in the world of genuine science. It is a waste of scarce resources that would be much better used in protecting the genuine archaeological heritage and is diverting attention from the pressing problems that are affecting professional archaeologists in Bosnia-Herzegovina on a daily basis.*

So clearly, we have battle lines drawn, and two sides each claiming to have the genuine science. Osmanagić is the one making the new extraordinary claim, so the burden of proof is on him. Let's point our skeptical eye at his evidence.

Osmanagić refers to Visočica as the Pyramid of the Sun. He says that it's only one of five such pyramids in the valley, the others being the Pyramid of the Moon, the Dragon, the Earth, and the Pyramid of Love. None of the others have any apparent pyramid shape to them, they are simply hills, so what tipped off Osmanagić that they are ancient pyramids?

Osmanagić founded the Archeological Park - Bosnian Pyramid of the Sun Foundation to solicit donations and fund his investigation, though it's mainly funded from his own fairly deep pockets. The Foundation commissioned one of its members, geophysicist Dr. Amer Smailbegovic to study available thermal images of the area, looking for thermal inertia. Basically, looking for areas that cool faster at night, and warm faster in the morning, than the surrounding area. Smailbegovic's report claims to have identified nine pyramids in this manner, five of them in Visočica's valley. Neither Smailbegovic nor Osmanagić offer any plausible explanation of why man-made pyramids might exhibit such a property, but it turns out there is a good reason that they do. The large pyramids in Egypt and Latin America are the highest points around, and a comparison of Smailbegovic's thermal images with topography maps show that (big surprise) it's generally the highest geographic points (hilltops, ridges, pyramids, whatever they are) that exhibit the lowest thermal inertia. If Smailbegovic did control for this fairly obvious correlation, he did not mention it in his report. When you choose your conclusion first, and then look for data that supports it, you're virtually guaranteed to get the results you want.

Osmanagić also offers as evidence the existence of a vast network of tunnels connecting all of his pyramids, and which exit at the summit of each. If true, this would certainly be interesting, but it would not, as he claims, make it consistent with man-made pyramids, as no known pyramids had any such network of tunnels connecting them that we know of. He offers video of one of the tunnels being explored. However this tunnel is the only one found so far, went back only 300 meters, and is located 3 kilometers away from Visočica. It could be an old mine shaft, nobody really knows (the area has been mined for coal, iron, and copper since Roman times); but there's certainly no evidence that the tunnel goes anywhere near Visočica or any other of Osmanagić's hills. Apparently he deduced the existence of a tunnel network from old, unsourced local stories about children going into one hill and emerging at the top of another. So much for the alleged "network of tunnels". If it exists at all, it certainly has never been found or detected.

Further evidence in favor of the pyramid comes from Harry Oldfield, an enthusiast in New Age energy crystals and aura photography. He took video of Visočica using a camera that digitally replaces colors, to which he gives the scientific-sounding name Polycontrast Interference Photography, and which he claims provides a "real time, moving image of the energy field." Technically, replacing colors just alters the visual image, it does not change the fact that the camera is capturing only visual data. Osmanagić, who refers to Oldfield as Dr. Oldfield for reasons known only to himself, analyzed this video and stated:

> ...The energy fields are vertical, as opposed to horizontal, which is the case with naturally occurring hills. In contrast to natural phenomena where the energy fields are fixed, these electromagnetic fields are pulsating and non-homogenous. The Bosnian Pyramid of the Sun is in fact acting like a giant energy accumulator which continually emits large quantities of energy. It is the proverbial perpetuum mobile, which got its start in the distant past and continues its activity without respite.

In Oldfield's color replacement video, brightness gradients in the sky appear as different colored bands, as is fairly obvious from a glance at the video. In explaining how he chose which colors to replace with which, Oldfield says "Some clairvoyants and mystics with their gifts helped me develop some of the filters in PIP which simulate what they see." If you understand what simple color replacement means, you should be able to judge for yourself the validity of Oldfield's video as proof that Visočica is a man-made pyramid.

Osmanagić's diggings on Visočica have been mainly to expose what appears to be extensive hand-laid stone roads, walls, and plazas, also including what appear to be ancient concrete blocks stacked in regular patterns. Although these structures look pretty convincing, geologists have yet to be impressed. The apparently nicely paved surfaces have been conclusively identified as natural strata. Local geologists call this series of strata the Lasva series, found throughout the region. This includes the blocks of conventional conglomerate, which Osmanagić misidentifies as artificial concrete. Similar structures that are equally or more impressive are found throughout the world.

Osmanagić does have his supporters, and it's not just he who identifies the conglomerate as concrete. These pronouncements claimed as scientific discoveries are made by the Board of Directors of Osmanagić's foundation, who are essentially the same people as the fifty or so who attended his ICBP 2008, the First International Conference on the Bosnian Pyramids. Occasionally the Foundation has named scientists from other countries as participants on their team. Osmanagić once said:

> Foreigners on our team include experienced archaeologists such as Richard Royce from Australia, Allyson McDavid from U.S., Chris Mundigler from Canada, Martin Aner from Austria.

Shall we take them one by one? Royce Richards (not Richard Royce) found that he had been named as "Senior Archaeologist" on the Foundation web site after merely sending Osmanagić an email, and describes the project as "snake oil" and "bollocks". Allyson McDavid does not participate in the project and is an illustrator, not an archaeologist. Chris Mundigler spent that year on a different archaeological project in a different country. There is no mention of various spellings of Martin Aner on the Foundation's web site that I could find.

The more I searched, the more people I found with similar stories. Sead Pilav and Grace Fegan are two more such victims. But I did find at least one case where a scientist named by the Foundation actually does exist and did visit the site for 45 days, a Dr. Ali Barakat from Egypt.

Dr. Barakat was reported to have been "sent by Cairo to assist Osmanagić's team." But when Egypt's Supreme Council for Antiquities was asked by *Archaeology Magazine* if this was true, the reply came directly from the Secretary General himself, Dr. Zahi Hawass:

> *The discoverer of the "pyramid" in Bosnia, Semir Osmanagić, who claims that a hill near the Bosnia River is a man-made structure built before the end of the last Ice Age, is not a specialist on pyramids. His previous claim that the Maya are from the Pleiades and Atlantis should be enough for any educated reader. This "pyramid" is actually a sloping hill near a village. ...Mr. Barakat, the Egyptian geologist working with Mr. Osmanagić, knows nothing about Egyptian pyramids. He was not sent by the SCA, and we do not support or concur with his statements.*

But it's worth pointing out that the scientific method does not permit us to use the business ethics of the Foundation as evidence that their scientific claims are not true. But they're certainly a red flag. And red flags turn out to be pretty much all Osmanagić has to justify his theory. The geology tells us that if the "pyramid" was man-made, it was constructed by laying the existing strata right back exactly where they were naturally formed, layer by layer, from bottom to top. This is one case where we can make a conclusion stronger than a lack of sufficient evidence to justify the claim. Visočica itself gives us all the physical, testable evidence needed to conclusively disprove the existence of the Bosnian Pyramid of the Sun.

References & Further Reading

Harding, Anthony. "The Great Bosnian Pyramid Scheme." *British Archaeology.* 1 Jan. 2007, Issue 92.

IRNA. "Geology of the Bosnian Pyramids." *Le site d'Irna.* IRNA, 25 Nov. 2006. Web. 10 Jan. 2010. <http://irna.lautre.net/Geology-of-the-Bosnian-pyramids.html>

Rose, Mark. "The Bosnia-Atlantis Connection." *Archaeology.* 27 Apr. 2006, Online Feature.

Schoch, Dr. Robert. "The Bosnian Pyramid Phenomenon." *The New Archaeology Review.* 1 Sep. 2006, Volume 1, Issue 8: Pages 16-17.

Woodard, Colin. "The Mystery of Bosnia's Ancient Pyramids." *Smithsonian Magazine*. Smithsonian Institution, 1 Dec. 2009. Web. 14 Jan. 2010. <http://www.smithsonianmag.com/history-archaeology/The-Mystery-of-Bosnias-Ancient-Pyramids.html>

25. The Lucifer Project

One of the most dramatic events in Arthur C. Clarke's *2001: A Space Odyssey* series of books happens in the second installment, *2010: Odyssey Two*. The strange alien monolith orbiting Jupiter somehow replicates itself billions of times, apparently using matter from Jupiter itself, condensing the gas giant down into a smaller, denser, hotter mass until it suddenly achieves sustained nuclear fusion. Thus, Jupiter fulfills the destiny denied it by nature as what some astronomers have termed a "failed star". This new star brings life to its constellation of Earth-like moons, becoming a solar system within a solar system, and is named Lucifer by the people of Earth, who henceforth have *two* suns in the sky.

> *Jupiter was not moving from its immemorial orbit, but it was doing something almost as impossible. It was shrinking — so swiftly that its edge was creeping across the field even as he focused upon it. At the same time the planet was brightening, from its dull gray to a pearly white. Surely, it was more brilliant than it had ever been in the long years that Man had observed it...*

Most of us consider this the stuff of science fiction; there are too many physical reasons why it couldn't actually happen, a few of which are raised by Clarke's characters as they witness the event incredulously. However, a number of conspiracy theorists (most of the variety who still believe in the Face on Mars) believe that this is not only possible, but that it is an actual project in the works at NASA. And, moreover, that what they term "The Lucifer Project" has already been attempted.

Details vary, the most significant of which is the confusion over whether Lucifer would involve Jupiter or Saturn. The *Space Odyssey* movies and books all use Jupiter, except for the original book, which was based on an early version of Clarke and Kubrick's screenplay that used Saturn (Saturn's rings later proved too great of a special effects challenge). Many of the modern conspiracy theories bring the story back around to Saturn as well, but really for all practical purposes we're talking about "a gas giant", think of either Saturn or Jupiter, whichever you please. Doesn't make much difference as far as reality is concerned.

The main element of these deep-space NASA missions that fuels the conspiracy is the RTGs, or radioisotope thermoelectric generators, that power space probes such as Cassini, Galileo, Voyager, and others. Past Mars there's not enough sunlight to provide the power a spacecraft needs, and so these RTGs are the only option we have. Russia has used similar generators to power about 150 lighthouses along its extremely remote northern coast. Heat from a radioactive element, usually plutonium-238, goes through a thermocouple, which is a material that produces a direct electrical current when heat is applied to it. RTGs have no moving parts and are extremely simple and reliable.

Believers in the Lucifer Project conjecture that such a payload of radioactive material would act like an atomic bomb in the high-pressure depths of a gas giant, and they suppose that this would somehow ignite the entire planet, turning the whole thing into a small star. This would act as a sun for its moons, turning them into habitable worlds. Saturn's moon Titan is usually cited, the claim being that NASA plans to turn it into a human colony for some unknown nefarious purpose.

In 2003, the Galileo spacecraft's mission was ended by deliberately crashing it into Jupiter, in order to absolutely avoid any possibility of contaminating Jupiter's moons with bacteria from Earth. A guy named Jacco van der Worp, now an advocate for the 2012 apocalypse, went on the Coast to Coast AM radio program and claimed that such a collision would cause the plutonium in the RTGs to immediately implode, triggering an atomic explosion. Richard Hoagland, the space conspiracy theorist who believes that many of the features found on Mars are ruins of ancient civilizations, and that NASA is covering it up, heard of van der Worp's idea and ate it up, claiming that a mysterious black spot that appeared briefly on Jupiter's surface a month later was evidence of this explosion. Hoagland asserts that Galileo would have broken up in Jupiter's atmosphere, and that it would have taken one month for the plutonium capsules to fall through Jupiter's increasingly dense atmosphere until such a pressure was reached that the capsules would implode. Hoagland concludes that the protection of Jupiter's moons from contamination was just a cover story for NASA's attempted creation of Lucifer.

The Cassini orbiter at Saturn is scheduled to terminate in 2012, however NASA has not yet decided whether to crash it into a smaller moon (where RTG contamination is not a problem) or to leave it in a high parking orbit. Saturn's rings make an approach for a Galileo-style crash into the gas giant too difficult.

There are a number of differences between an RTG and an atomic or thermonuclear warhead. The grade of plutonium is one difference. The RTG uses reactor grade plutonium, while a weapon uses weapon-grade plutonium. The difference is that weapon grade contains less than 7% Pu-240. Reactor grade has more. Not only does this make a chain reaction more difficult to sustain, it also makes the material more radioactive and more difficult to work with and store. In 1977, the United States declassified a 1962 underground nuclear test at the Nevada test site in which non-weapon grade plutonium was used. Although the explosive yield was quite low, the test proved that the plutonium grade alone doesn't disprove the Lucifer conjecture.

But the main reason that an RTG could not explode like a weapon is its structure. Each of Cassini's three RTGs contains 72 marshmallow-sized pellets of plutonium, each weighing about 150 grams, and each separately enclosed in iridium inside a shock-proof graphite impact shell. Four of each of these are enclosed within one of 18 separate General Purpose Heat Shell modules, each with its own separate heat shield and impact shell. Should any kind of crash or problem happen, including breaking up during a re-entry, these impact shells separate from each other and scatter.

Conversely, in order to detonate Pu-238, you need a single critical mass of solid plutonium weighing at least 10 kg. This critical mass has to be imploded with a simultaneous explosion from all sides, applying sudden pressure precisely from all angles at the same exact instant. Obviously this couldn't happen with an RTG design. Although each RTG does theoretically have enough plutonium to make up a critical mass, there isn't any way that it could all be brought together into the right shape. An implosion triggered atomic device needs to have its critical mass in a very specific configuration. Any type of pressure or crash event has already sent all the separate impact shells scattering about space, and each is far too small to ever achieve critical mass and implode. No way, no how, physics simply do not make it possible for a chunk of less than critical mass to initiate a chain reaction, no matter what environment it's put in.

Proponents of the NASA conspiracy state that the high pressure of the deep atmosphere inside a gas giant will provide the implosion pressure, but they do not offer a solution for the critical mass problem. I searched and searched, and the best document I could find by conspiracy theorists, by an anonymous author, admits that the pellets are 150 grams but states that plutonium-238 requires only 200 grams to reach critical mass. This is simply wrong, but even if it were true, 150 is still less than 200. However the author seems to simply ignore this, skips over it, and says that a 600 kiloton explosion would result.

So let's grant that an RTG could somehow result in a 600 kiloton atomic explosion on a gas giant. This is only a tiny fraction of the firepower of some of the thermonuclear tests done on Earth, the largest being the Soviet Union's 1961 *Tsar Bomba* shot with a yield of 50 megatons. That didn't turn the Earth into a small sun, it was a barely visible pinprick on our gigantic planet. So why would this far, far smaller explosion have such a drastic effect on a gas giant? Well, like our sun, the gas giants are composed largely of hydrogen and helium. In the intensely confined pressure inside an atomic explosion, fusion happens among these elements and causes the runaway thermonuclear chain reaction. In a nuclear explosion on Earth, this chain reaction quickly runs out, because of a lack of pressure and fuel. But inside the sun, there is tremendous fuel available and tremendous pressure from the sun's powerful, crushing gravity. This is called gravitational confinement, and it's the reason the sun's nuclear reaction is ongoing.

Stars that are less massive than the sun have less gravity. Beyond a certain limit, they have inadequate gravitational confinement. These are called brown dwarfs. Because of their density and gravity, all brown dwarfs happen to be about the same physical size as Jupiter. However their mass ranges from 1 to about 90 Jupiter masses. Above this limit, they would have adequate gravitational confinement and could sustain fusion. But inside this range, at which Jupiter is at the extreme lowest end, they don't and can't. Some astronomers don't make a clear distinction between what constitutes a gas giant and what constitutes a brown dwarf, but one feature they share is mass that's way too low for sustaining fusion. An atomic or even thermonuclear explosion inside Jupiter would fizzle out the same way it does on Earth. Saturn, with less than a third of Jupiter's mass, is even farther from achieving gravitational confinement.

Even Arthur C. Clarke didn't pretend that his fiction was plausible. At Lucifer's ignition, one of Clarke's Russian scientists, Vasili, said:

Oh, I can see a dozen objections — how would they get past the iron minimum; what about radiative transfer; Chandrasekhar's limit.

And like Vasili, we've only touched upon a couple of the problems, but certainly among the most intractable for those who believe that our tiny little space probes are the harbingers of planetary death and new solar systems. Enjoy your science fiction stories and enjoy the science coming back through Cassini's telemetry, but please don't confuse the two. Saturn and Jupiter are here to stay.

REFERENCES & FURTHER READING

Beatty, J., Petersen, C., Chaikin, A. *The new solar system 4th Edition.* Cambridge: Sky Publishing Corporation, 1999. 194-195.

Blanchard, A. et al. *Updated Critical Mass Estimates for Plutonium-238.* Aiken, SC: Westinghouse Savannah River Company, 1999.

Griffin, Michael D.; French, James R. *Space Vehicle Design Second Edition.* Reston: American Institute of Aeronautics and Astronautics, Inc, 2004. 497-500.

NASA. "Spacecraft Power for Cassini." *NASA Fact Sheet.* NASA, 1 Jul. 1999. Web. 16 Jan. 2010.
<http://georgenet.net/hubble/cassini_pdf/power.pdf>

O'Neill, Ian. "Project Lucifer: Will Cassini Turn Saturn into a Second Sun? (Part 1)." *Universe Today.* Universe Today, 24 Jul. 2008. Web. 15 Jan. 2010.
<http://www.universetoday.com/2008/07/24/project-lucifer-will-cassini-turn-saturn-into-a-second-sun-part-1/>

Planning & Human Systems, Inc. *Atomic Power In Space: A History.* Washingon, DC: U.S. Department of Energy, 1987.

26. FEMA Prison Camps

Today we're going to have a look at more than 800 sites inside the United States, said by some to be prisons operated by FEMA (the Federal Emergency Management Agency), for the purpose of holding as many as ten million American citizens prisoner, with no criminal charges filed. YouTube carries videos of such empty prison compounds, and dozens of web sites like libertyforlife.com and abovetopsecret.com showcase photographs and reports by independent investigators. These alleged prisons are hidden everywhere in plain sight: A train yard or shipping terminal mysteriously surrounded by barbed wire; a closed military base with some new construction happening; or even just a vacant site that seems like it could be a good location. A paranoid conspiracy theory, you say? Perhaps. But might there actually be reason to believe that plans for just this scenario really do exist?

When I first heard the FEMA Prison Camp conspiracy story, it seemed ridiculous and paranoid at face value. But when I finally dug in to research it, I started by searching for the origins of the rumors, and found to my surprise that nearly all of the legal foundation and precedent for such a plan does in fact exist. One primary source of fuel for the fire is Garden Plot, the Department of Defense's civil disturbance plan to assist local authorities during times of civil unrest, natural disasters, or other emergencies. Garden Plot has been notably activated after 9/11, and also during the 1992 Los Angeles riots. Garden Plot's authority comes from Article I of the United States Constitution, that states in part "Congress shall have power... to provide for calling forth the Militia to execute laws of the Union, suppress Insurrections, and repel Invasions." Taking it one step further, Title 10, sections 331-334 of federal law authorize the President to suppress insurrections, rebellions, and domestic violence by executive order. In recent years, the Insurrection Act and the Posse Comitatus Act, which limited executive powers to deploy the military for purposes of law enforcement, have been amended to broaden the scope of exceptions.

Periodically, the government conducts interagency readiness exercises to prepare for such contingencies. Past examples of such exercises that are frequently cited by Prison Camp conspiracy guys include Rex-82 *Proud Saber* (where Rex-82 stands for Readiness Exercise, 1982), and Rex-84 *Night Train*. In these exercises, agencies deal with such problems as major strikes and unlawful assemblies. They may impose martial law, arrest large numbers of people, and handle mass relocation of civilians. Now, obviously, there are times when this is a good thing and we want it to happen. A lot of us were glad that the National Guard came in to help quell the L.A. riots. If a large force of Oklahoma City style militiamen started blowing up a city, I'd want the Army to have pretty extensive powers to put a stop to it. But, unfortunately, everything is a double-edged sword. For the government to have this level of power when it's needed, it means they also have it when they *say* it's needed, and when you and I might not necessarily be in favor of it. What if Congress suddenly made all guns illegal? They would sure as hell need prison camps to hold ten million rioting citizens. Such powers necessarily do exist.

So there's our legal authority and precedent. If FEMA decides they want 800 prison camps maintained in readiness, they have the legal right to do so, and to use them if ordered. But do they? Do they actually anticipate this need to the point that there really are 800 manned prison camps ready to go? That's another story.

Despite the reasonable plausibility, most of the sources trumpeting FEMA prison camps are clearly delusional, blaming the "Illuminati" for the prison camps, and claiming that they are to support the "new world order". One online list of prison locations describes a Mojave Desert location as a "Fully staffed full gassing/cremating death camp with airstrip, dedicated to the termination red/blue list under martial law," and states that it was recently toured by "high level Illuminati Luciferians". A camp in Alaska is said to have a capacity of two million, despite Alaska's total population of less than 700,000.

And now that President Obama is in office, the charges now fall onto his shoulders. One report says that Obama has already ordered the immediate opening of America's "vast gulag of concentration camps" to handle the "social upheaval" caused by the "economic collapse", and is in the process of incarcerating 775,000 citizens as part of "Project ENDGAME". Fear that this mass arrest and execution of Americans is already in progress is commonly expressed on a lot of these web sites. You might remember former U.S. Representative and Green Party presidential candidate Cynthia McKinney, perhaps best known for her 2006 punching of a Capitol police officer who stopped her when she refused to show proper credentials while bypassing the Capitol's metal detectors. Following Hurricane Katrina, she announced that 5,000 New Orleans prisoners had been executed by the military:

> *Her son's charge by the Department of Defense was to process 5,000 bodies that had received a single bullet wound to the head, and these were mostly males... The data about these individuals was entered into a Pentagon computer. And then reportedly the bodies were dumped in the swamp in Louisiana... I have verification from insiders, who wish to remain anonymous, at the Red Cross, that this is true.*

Because clearly, the Red Cross would be deeply involved in such a plot.

McKinney was also the driving force behind the 2005 Congressional briefing where she presented testimony and numerous witnesses claiming that the U.S. was behind the 9/11 attacks. In fact, when you search the Internet for FEMA prison camps, you find a lot of the information is deeply interwoven with 9/11 conspiracy theories. The same familiar names appear: Like when Halliburton announced its fourth quarter results for 2005, one of the line items was a $385 million five-year contract to support temporary detention capabilities for Immigration and Customs Enforcement. Conspiracy theorists jumped on this and said it was proof that the prisons are being built. I don't know how many prisons you think can be built, staffed, and operated for $77 million a year, but our local megachurch here has spent more than that alone just building a parking garage and a bookstore cafe.

Now of course, we do know that the U.S. government has, in the past, set up prison camps into which law-abiding American citizens have been forcibly relocated, by Presidential order. This happened between 1942 and 1945, when people of Japanese ancestry were rounded up and placed into concentration camps to prevent them from aiding the Japanese during World War II. I recently visited Manzanar, the best preserved of these sites, which is now operated by the National Park Service. You can tour the grounds and they have a great museum, with very frank exhibits that make no excuses for what happened. There is no secrecy or coverup about it. Even today, the U.S. government operates detention facilities around the world, like Guantanamo Bay, which is well known and fully disclosed. Garden Plot and the readiness exercises are all matters of public record.

But there are other sites, such as Diego Garcia Island in the Indian Ocean, for which reasonable evidence exists that it's been used as a detention facility, but that no official admission is made. In fact, it's denied. Whether true or not, the existence of Diego Garcia would not constitute good evidence that concentration camps exist inside the United States; there are significant differences. Whatever might happen on Diego Garcia happens behind closed doors, whereas the claimed concentration camps are right out in broad daylight. And if you're going to illegally rough up a terrorism suspect, a small room overseas is a great place to do it; but a wide-open concentration camp in the public eye, designed for tens of thousands, would be a rather poor choice. So, to a responsible skeptic, the other examples of government behavior that we have constitute pretty poor support for the existence of domestic concentration camps.

At the time Japanese Americans were rounded up and put into Manzanar, it was a reasonable precaution according to the standards of the times. I used to find it hard to believe that people could think so differently only a few decades ago, but in 1999 I bought a house in a neighborhood laid out in the 1930's, and in the original deed was a clause that the buyer may not to sell the house to "a Negro, a Jew, or a Chinaman." Times change. Manzanar would never happen today, and although the government technically has the same powers to do it now as it did then, it's not realistic to be concerned about it recurring.

Our man Benjamin Franklin famously said (though somewhat apocryphally): "Any society that would give up a little liberty to gain a little security will deserve neither, and will lose both." Having to take off your shoes to get on a plane and not being allowed to have a water bottle are infringements that Ben would have deemed unacceptable first steps, but they are also a far cry from millions of civilians being thrown into prison camps. We choose to elect politicians who don't want us to bring water bottles onto planes, because (for better or for worse) that's what's important to our society right now. I don't remember anyone electing a politician who wants to throw millions of Americans into prison camps. To make effective electoral decisions, you need to maintain a healthy skepticism, and not go off the deep end and suppose that every Halliburton contract is a slippery slope leading to Americans being gassed in military concentration camps. If you see barbed wire around a train yard, consider the possibility of other explanations (like the train company doesn't need stuff being stolen) before you conclude that the Illuminati are out to kill you.

References & Further Reading

Aaronovitch, D. *Voodoo History: The Role of the Conspiracy Theory in Modern History.* New York: Riverhead, 2010.

Arkin, Wiliam M. *Encyclopedia of the U.S. Military.* New York: Harper & Row, 1990. 574-575.

Barkun, M. *A culture of conspiracy: Apocalyptic visions in contemporary America.* Berkeley, Ca.: University of California Press, 2003. 73-74.

Irons, Peter. *Justice At War.* New York: Oxford University Press, 1983. 25-47.

Jacobs, J. *Socio-legal foundations of civil-military relations.* New Brunswick, NJ: Transaction, Inc, 1986. 51-62.

27. How Old Is the Mount St. Helens Lava Dome?

Today we're going to point our skeptical eye at one of the key players in the debate between geologists and Young Earthers over the age of the Earth. In June of 1992, Dr. Steven Austin took a sample of dacite from the new lava dome inside Mount St. Helens, the volcano in Washington state. The dacite sample was known to have been formed from a 1986 magma flow, and so its actual age was an established fact. Dr. Austin submitted the sample for radiometric dating to an independent laboratory in Cambridge, Massachusetts. The results came back dating the rock to 350,000 years old, with certain compounds within it as old as 2.8 million years. Dr. Austin's conclusion is that radiometric dating is uselessly unreliable. Critics found that Dr. Austin chose a dating technique that is inappropriate for the sample tested, and charged that he deliberately used the wrong experiment in order to promote the idea that science fails to show that the Earth is older than the Bible claims. Yet the experiment remains as one of the cornerstones of the Young Earth movement.

Of most people who have heard of this incident before, that's probably about the total depth of what they've heard. And there's pretty good reason for this: Geology dating is pretty complicated, and if you look at Dr. Austin's paper or at any scholarly criticism of it, your eyes will quickly glaze over from the extraordinary detail and intricacy. So I thought this would be a great place to point our skeptical eye, and see how much of the chaff we can cut through to see what the bare facts of the case really are. Obviously both sides of this debate have agendas to promote, and that means that any summary you're likely to read was probably motivated by one agenda or the other.

Let's begin with a basic understanding of the radiometric dating technique used, K-Ar, or potassium-argon. This dating technique depends on the fact that the radioactive isotope of potassium, ^{40}K, naturally decays into other elements, as do all unstable radioactive elements. There are two ways that this happens to ^{40}K. About 89 percent of the time, a neutron inside the ^{40}K undergoes beta decay, in which the neutron decays into a proton and an electron. This gain of a proton turns the potassium into calcium. But about 11 percent of the time, an extra proton inside the ^{40}K captures one of its electrons and merges with it, turning the proton into a neutron and a neutrino, and converting the potassium into argon. In both events, the atomic mass remains unchanged, but the number of protons changes, thus turning the element from one to another. This happens to ^{40}K everywhere in the universe that it exists, and at the same rate, which is a half-life of 1.2 billion years. This means that if you have 1000 atoms of ^{40}K, 1.2 billion years later you'll have 500, and 1.2 billion years after that you'll have 250. You'll also have 83 argon atoms, and 667 calcium atoms. If I take a sample and measure an argon to potassium ratio of 83:250, I know that this sample is 2.4 billion years old.

However, all of these numbers are probabilities, not absolutes. You need to have a statistically meaningful amount of argon before your result would be considered significant. Below about 10,000 years, potassium-argon results are not significant; there's not yet enough argon created. The 11% of the time that potassium decays into argon and not calcium is also a probability, so this contributes to the result having a known margin of error. In addition, the initial amount of ^{40}K that you started with is never measured directly; instead, it is assumed to always be .0117% of the total potassium present, which is the known distribution in nature. This has a standard deviation, so it also contributes to the margin of error. So when my result says the sample was 2.4 billion years old, this is only correct if the sample was at least 10,000 years old to begin with, and it's only correct plus or minus a calculated margin of error, in this example about 600,000 years. The bell curve of probable age starts at about 1.8 billion years, peaks at 2.4 billion, and dips back to the baseline at 3 billion. So whether you call it an exact science or not is a matter of linguistics. Although the exact age can't be known, the probabilities can be exactly calculated.

Since Dr. Austin's sample was known to have solidified in 1986, its argon content was clearly well below the threshhold where an amount of argon sufficiently useful for dating could have been present. And even that threshhold applies to only the most sensitive detection equipment. Potassium-argon dating is done by destructively crushing and heating the sample and spectrally analyzing the resulting gases. The equipment in use at the time at the lab employed by Dr. Austin, Geocron Laboratories, was of a type sensitive enough to only detect higher concentrations of argon gas. Geocron clearly stated that their equipment was only capable of accurate results when the sample contained a concentration of argon high enough to be consistent with 2,000,000 years or older.

And so, by any standard, it was scientifically meaningless for Dr. Austin to apply Geocron's potassium-argon dating to his sample of dacite known to be only six years old. But let's ask the obvious question. If there wasn't yet enough argon in the rock to be detectable, and the equipment that was used was not sensitive enough to detect any argon, how was enough argon found that such old results were returned?

There are two possible reasons that the old dates were returned. The first has to do with the reason Geocron's equipment was considered useful only for high concentrations of argon. There would always be a certain amount of argon inside the mass spectrometer left over from previous experiments. If the sample being tested is old enough to have significant argon, this leftover contamination would be statistically insignificant; so this was OK for Geocron's normal purposes. But for a sample with little or no argon, it would produce a falsely old result. This was undoubtedly a factor in Dr. Austin's results.

The second possibility is that so-called "excess argon" could have become trapped in the Mount St. Helens magma. This is where we find the bulk of the confusing complexity in Austin's paper and in those of his critics. The papers all go into great detail describing the various ways that argon-containing compounds can be incorporated into magma. These include the occlusion of xenoliths and xenocrysts, which are basically contaminants from existing old rocks that get mixed in with the magma; and phenocrysts, which are crystals of all sorts of different minerals that form inside the rock in different ways depending on how quickly the magma cools. 95% of these papers are geological jargon that will make your head spin: Page after page of chemical compositions, mineral breakdowns, charts and graphs, and all sorts of discussion of practically every last molecule found in the Mount St. Helens dacite.

Summarizing both arguments, Dr. Austin claims that xenoliths and xenocrysts were completely removed from the samples before testing, and that the wrong results are due to phenocrysts, which form to varying degrees in all magma, and thus effectively cast doubt on all potassium-argon testing done throughout the world. It's important to note that his arguments are cogent and are based on sound geology, and are often mischaracterized by skeptics. He did not simply use the wrong kind of radiometric dating as an ignorant blunder. He was deliberately trying to illustrate that even a brand-new rock would show an ancient age, even when potassium-argon dating was properly used.

Austin's critics charge that he ignored the probable likelihood that the limitations of Geochron's equipment accounts for the results, just as Geochron warned. They also charge that he likely did not remove all the xenoliths and xenocrysts from his samples. However, neither possibility can be known for sure. Certainly there is no doubt that the test was far outside the useful parameters of potassium-argon dating, but whereas critics say this invalidates the results, Austin concludes that his results certify that the test is universally useless.

If we allow both sides to have their say, and do not bring a bias preconditioning us to accept whatever one side says and to look only for flaws in the other side, a fair conclusion to make is that both sides make valid points. Austin does indeed identify a real potential weakness in potassium-argon dating. However he is wrong that his phenocrysts constitute a fatal flaw in potassium-argon dating previously unknown to geology. In fact, the implications of phenocrysts were already well understood. Yes they are one of the variables, and yes, in some samples they do push the error bars. However, the errors they introduce are in the range of a standard deviation, they are not nearly adequate to explain errors as gross as three or more orders of magnitude, which would be necessary to explain the discrepancy between the measured age of rocks and the Biblical age of the Earth.

Such variables are also a principal reason that geologists never rely on just one dating method, with no checks or balances. That would be pretty reckless. For most rocks, multiple types of radiometric dating are appropriate; and in practice, multiple samples would always be tested, not just one like Austin used. In combination, these tests give a far more complete and accurate picture of a rock's true age than just a single potassium-argon test could. In addition, stratigraphic and paleomagnetic data can often contribute to the picture as well. From many decades of such experience, geologists have excellent data that guides proper usage of each of these tools, and they don't include gross misuse of potassium-argon dating.

What Austin did was to exploit a known caveat in radiometric dating; dramatically illustrate it with a high-profile test using the public's favorite volcano, Mount St. Helens; and sensationalize the results in a paper that introduces nothing new to geologists, but that impresses laypeople with its detailed scientific language. Occasionally scientists do actually make huge discoveries that everyone else in their field had always missed, but such claims are wrong far more often than they're right;

and Dr. Austin and his finding that radiometric dating has always been useless is a perfect example.

REFERENCES & FURTHER READING

Austin, Steven A. "Excess Argon within Mineral Concentrates from the New Dacite Lava Dome at Mount St. Helens Volcano." *Creation Ex Nihilo Technical Journal.* 1 Dec. 1996, Volume 10, Number 3: 335–343.

Dalrymple, G.B. "Radiometeric Dating Does Work!" *Reports of the National Center for Science Education.* 1 May 2000, Volume 20, Number 3: 14-19.

Henke, Kevin R. "Young-Earth Creationist 'Dating' of a Mt. St. Helens Dacite: The Failure of Austin and Swenson to Recognize Obviously Ancient Minerals." *No Answers in Genesis!* Australian Skeptics Inc., 21 Jul. 2007. Web. 24 Dec. 2009. <http://noanswersingenesis.org.au/mt_st_helens_dacite_kh.htm>

Walker, Mike. *Quaternary Dating Methods.* West Sussex, U.K.: John Wiley & Sons, Ltd., 2005. 58-62.

Wiens, Roger C. "Radiometric Dating A Christian Perspective." *Science in Christian Perspective.* American Scientific Affiliation, 1 Jan. 2002. Web. 24 Dec. 2009. <http://www.asa3.org/aSA/resources/wiens2002.pdf>

28. Binaural Beats: Digital Drugs

Today we're going to put on our headphones, kick back in the beanbag, and get mellow to the soothing sounds of the latest digital drug: binaural beats. These computer generated sound files are said to massage your brain and produce all sorts of effects, everything from psychedelic experiences to behavior modification. Let's point our skeptical eye at the science of binaural beats, and especially at some of the claims made for them.

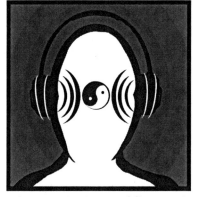

A binaural beat is created by playing a different tone in each ear, and the interference pattern between the slightly differing frequencies creates the illusion of a beat. It's intended to be heard through headphones, so there's no cross-channel bleed across both ears. There isn't actually any beat, it's just an acoustic illusion:

If you search the Internet for "binaural beats" you'll quickly find there's a whole industry built on the idea that listening to binaural beats can produce all kinds of desired effects in your brain. It can alter your mood, help you follow a diet or stop smoking, get you pumped up for a competition, calm you down, put you to sleep, enhance your memory, act as an aphrodisiac, cure headaches, and even balance your chakra. Binaural-Beats.com offers a $30 CD that they call the world's first "digital drug". They claim it can get you drunk without the side effects. I-Doser.com offers a range of music tracks that they say simulates a variety of actual pharmaceuticals, such as Demerol, Oxycontin, and Vicodin. Suffice it to say that no matter what superpower you're looking for, someone on the Internet sells a binaural beat audio file claimed to provide it.

You don't have to buy one, though. It's not too hard to make your own binaural beat, and free software is widely available to do just that. A binaural beat consists of two simple tones, and most people add background pink noise. Nothing special.

But the question is: Does it have a special effect on the brain? A lot of people think so. The basic claim being made for binaural beats is "resonant entrainment". Entrainment, in physics, is when two systems which oscillate at different frequencies independently are brought together, they synchronize with one another, at whatever the combined system's resonant frequency is. Examples of entrainment occur in animals in nature; for example the chirping of crickets or the croaking of frogs. Synchronization of menstrual cycles in women is another example. Even people coming together and dancing with one another is a type of entrainment. The basic claim for binaural beats is that the perceived low-frequency beat will entrain your brain wave pattern, thus forcing your brain into some desired state.

Most of these web sites give some brief explanation of entrainment. The example you hear most often is that of Dutch polymath Christiaan Huygens, who in 1665, hung two pendulum clocks next to each other on a wall. He noticed that the pendulums eventually matched each others' frequency, but always in antiphase, opposite to each other, as if canceling each other out. He'd try disturbing one or setting them in sync, but they'd always return to the same antiphase synchronization. Huygen's experience is widely touted on binaural beat websites as a demonstration of how systems can become spiritually connected through some energy field. However, they misunderstand what happened, and have not read the full story. Huygens also tried taking one clock off the wall, and as soon as they were no longer physically connected to one another via the actual wall, the effect disappeared. It was not the proximity of the clocks to one another that created the entrainment; it was their physical, mechanical connection to one another. As each pendulum swung it imparted an infinitesimal equal and opposite reaction to the wall itself. With two clocks on the wall, the system naturally sought the lowest energy level, according to the laws of thermodynamics; and both pendulums would thus swing exactly counter to each other, minimizing the system's total energy.

So to summarize their claim, they're saying that entrainment means that a binaural beat will cause your brain's electroencephalogram to match the pattern of the phantom beat. Well, if it did, entrainment certainly doesn't apply and would not be part of the equation, so we can scratch that off the list. But it doesn't make the claimed observation wrong. We do know that certain electroencephalogram waveforms are often associated with certain kinds of activity. For example, physical activity or REM sleep often produces an electroencephalogram with a sine wave of between 4 and 8 Hz, which we term a theta pattern. Waking relaxation with eyes closed often produces a pattern from 8 to 12 Hz, which is called an alpha pattern. There are only a few characterized patterns, and pretty general descriptions of what kinds of activities go with them. The claim made by the binaural beat sellers depends on much more granular and specific matches. For example, the claim that a binaural beat with a frequency of X produces the same effect in your brain as Vicodin is wholly implausible. Such claims presume that we know the exact frequency of the electroencephalogram in each of these desired conditions, and the fact is that brain waves don't work that way. It is wholly and absolutely implausible to say that desired brain condition X will occur if we get your EEG to read exactly X Hz.

Not only that, binaural beats presume that brain waves work in the opposite way that they do. Certain brain states produce certain brain waves; brain waves don't produce brain states. You just don't turn a dial to 6.5 Hz and induce instant happiness.

And so, while the claimed science behind binaural beats is unfounded, this doesn't mean that the effect isn't real and simply unexplained. Maybe you can listen to a certain binaural beat and induce a desired state, but for reasons we don't yet understand. So let's take a look at the research, and see if such an effect has actually been observed.

A 2008 study at Hofstra University played two different binaural beats and a control sound (a babbling brook) to patients with high blood pressure. There was no difference between the groups. In one small study from Japan that was published in the Journal of Neurophysiology in 2006, they played various binaural beats to nine subjects, and observed the resulting EEGs. They found great variability in the results. Their conclusion was that listening to binaural beats can produce activity on the human cerebral cortex, however the cause was more likely a conscious auditory reaction and was not correlated to the frequency of the binaural beat. However, a 2005 study published in Clinical Neurophysiology found that they were able to induce a desired frequency in the EEG matching the phantom beat frequency encoded in a binaural beat, however this was with a single subject and was neither blinded nor controlled.

But we don't need any studies to tell us that different people can listen to different kinds of music and be affected. A lot of people who work out have a workout playlist on their iPod that keeps them energized. Some people listen to certain music to help them fall asleep. The Muzak company has built an industry on relaxing music that will keep people in the mood to shop. Music does affect our mood, and so we already have every reason to expect binaural beat recordings to produce the same effect. Different people may find certain binaural beats to be relaxing or energizing. But, we've never found any reliable indication that a binaural beat's connection to our brain is any deeper or more meaningful than any other music track. We do know for a reasonable certainty that specific claims made by most sellers of binaural beats are not credible, and that there is no reason to think that the effect they're claimed to produce will work for you.

Well, except for one reason: The power of suggestion. If I give you a music track and tell you that it will cure your headache, you're more likely to report that it cured your headache than you are to say "Well it didn't effect my headache, but it made my short-term memory better." An interesting experiment would be to buy a binaural track claimed to induce drunkenness, for example; play it for five friends without telling them the claim, and then ask how it made each of them feel. Give them multiple choices to select from. Chances are they're going to respond all over the map. If you have a friend who is a believer in binaural beats, I suggest going ahead and setting up this little test.

So, in summary, binaural beats certainly do not work the way the sellers claim, but there's no reason to think they're any less effective than any other music track you might listen to that effects you in a way you like. If they make you sleepy (like they all do for me), use them to go to sleep. If they relax you or get you amped, use them for that. But don't expect them to be any more effective than regular music. If someone you know claims that they are, put them to the test, and bust the myth.

REFERENCES & FURTHER READING

Carter, C. "Healthcare performance and the effects of the binaural beats on human blood pressure and heart rate." *Journal of Hospital Marketing and Public Relations.* 1 Aug. 2008, Volume 18, Number 2: 213-219.

Karino, S., Yumoto, M., Itoh, K., Uno, A., Yamakawa, K., Sekimoto, S., Kaga, K. "Neuromagnetic responses to binaural beat in human cerebral cortex." *Journal of Neurophysiology.* 21 Jun. 2006, Volume 96, Number 4: 1927-38.

Padmanabhan, R., Hildreth, A.J., Laws, D. "A prospective, randomised, controlled study examining binaural beat audio and pre-operative anxiety in patients undergoing general anaesthesia for day case surgery." *Anaesthesia.* 7 Jul. 2005, Volume 60, Number 9: 874-877.

Pratt H., Starr A., Michalewski H.J., Dimitrijevic A., Bleich N., Mittelman N. "Cortical evoked potentials to an auditory illusion: Binaural beats." *Clinical Neurophysiology.* 1 Aug. 2009, 120, 8: 1514-1524.

Schwarz, D.W., Taylor, P. "Human auditory steady state responses to binaural and monaural beats." *Clinical Neurophysiology.* 1 Mar. 2005, Volume 113, Number 3: 658-668.

29. Coral Castle

In Homestead, Florida stands an oddity that is a favorite of paranormal enthusiasts: A sort of artificial rock park popularly called Coral Castle, though its official name is Rock Gate Park. Built from a total of 1,100 tons of locally quarried coral, the park is filled with astrologically themed structures like crescent moons, stars and planets, a sundial, obelisks, and a viewing port called the Telescope. The whole park is surrounded by an 8-foot wall of coral blocks weighing 15 tons each. The largest single block in the park weighs 30 tons, and two of the monoliths are 25 feet high. But the remarkable thing about Rock Gate Park is that it was built by a single man. Over a period of 28 years, from 1923 until his death in 1951, the diminutive 5-foot-tall Edward Leedskalnin, a Latvian, worked all by himself to quarry, cut, move, and position every single coral block. Coral Castle is almost always described as a mystery; a mystery that cannot be explained.

But that's where the story of Coral Castle only *starts* to get weird. Ed wasn't just interested in moving heavy blocks of coral, he also dabbled in electricity and magnetism. He built a hand-cranked electrical generator and wrote a number of short pamphlets, which he sold through local newspapers, on subjects as diverse as his own notions about magnetic current and the magnetic nature of life, as well as a political rant that sounds like it could have been written by Hitler. I've read his pamphlets on magnetism (they've available online) and to me they're quite childish, but perpetual motion enthusiasts and vortex energy believers have latched onto Ed as a great genius. One guy has even self-published an e-book called *Coral Castle Code* in which he believes Ed's writings unlock all the secrets of energy, life, and the true nature of the atom. They believe that Ed built Coral Castle using his unique insights into magnetism and electricity, perhaps levitating the blocks into place using magnetic vortices (whatever that means).

Very few people ever watched Ed actually do his work moving the blocks — more about this later — and so some believe that he deliberately worked in great secrecy to protect some presumed secret. Did Ed use some advanced technology to build Coral Castle? The first question a skeptic should ask is "Is it necessary to invoke a mystical or undiscovered technology as the only possible explanation?"

Meet Wally Wallington, a retired construction worker in Michigan. He's building a Stonehenge replica in his back yard, and he's using only levers and fulcrums. His equipment consists of sticks and stones. No wheels, no cranes, no pulleys, no metal or machinery. His favorite tool, and best ally, is gravity. There's a YouTube video of Wally raising a 19,200-lb concrete beam up into the air and tilting it up onto its end, just like at Stonehenge. He can also raise these beams horizontally into the air as high as you'd like, with little apparent effort. It's simply good old fashioned human ingenuity. All he does is lever the block enough to slide a fulcrum under its center, turning it into a giant teeter totter. Weight down the east end, and slide more wood in just west of the center. Move the weight to the west end, transferring the mass onto the west end of the fulcrum, and slide in more wood just east of the center. Back and forth, back and forth, going higher each time, rocking its way into the sky.

Wally takes a 1600-lb block and rolls it over the ground as easily as you'd roll a basketball. How? Imagine viewing the square end of a rectangular block. Now superimpose a circle over that square end, and imagine that circle rolling along. Visualize the path that the square would trace over the ground. Wally simply built a wooden roadway shaped like that trace — it looks like a line of half-circles placed next to each other, round side up — and he can give a 1600-lb block a shove and send it rolling down that path. It rolls smoothly about its center, just as if it *were* round.

Wally can also move a one-ton block 300 feet an hour by levering it up onto a pebble as a fulcrum, and walking it around onto another pebble a foot or so away.

Watching Wally move his blocks around looks so simple, the techniques seem so obvious, and yet I didn't think of them. I slapped myself on the forehead when I first watched his video. And, logically, I have to force myself to accept that there are *other* techniques, just as easy and just as inventive, that I also haven't thought of. Human beings have an annoying tendency to project our own ignorance. We tend to say "I can't think of a way to do this, therefore you can't either." As a kid I spent years with my LEGO kit trying to reinvent the automobile differential. I knew they existed, but I didn't know what the mechanism was. I finally gave up, and actually forced myself to conclude they were mechanically impossible, in dead opposition to the evidence. Then one day I finally saw one, and it was like watching Wally Wallington's video for the first time. I turned red as I realized I simply hadn't considered enough possibilities. I hadn't been innovative enough. Watching a simple, elegant solution to a problem we had ourselves considered insoluble can be a bitter pill to take, or it can be like opening a Christmas present and finding the perfect gift. It depends on your attitude.

It depends on whether you want there to be a solution, or whether you insist on a supernatural explanation and are predisposed to antagonism toward any notion that suggests otherwise.

We don't know whether Ed Leedskalnin used the same techniques as Wally Wallington, but we clearly have to acknowledge that there are ways to do it without machinery. When asked how he built his castle, Ed would answer "It's not difficult if you know how," and according to Coral Castle's web site, he said he was able to move the heavy blocks because he "understood the laws of weight and leverage well." Sounds to me like he figured things out the same way Wallington did.

Ed actually had a little more leeway than Wally, because Wally chose to use only technologies known to the ancient Britons, while Ed imposed no such restrictions on himself. Remember how I said very few people actually watched Ed work? Well, some did, and some brought cameras. There are photographs of Ed at work on his castle, lifting his blocks with a large tripod made of telephone poles perhaps 25 feet tall, using chains and a block and tackle system. When he once disassembled and relocated Rock Gate Park a few miles from Florida City to Homestead to escape an encroaching subdivision, the blocks were moved on a flatbed trailer towed behind a rented tractor. But his use of tripod cranes and tractors don't seem to fit in very well with the magnetic energy vortex theories, and so you won't find references to these pictures on most Coral Castle web sites.

A Wikipedia author conceded the existence of the photographs of Ed with his boringly Earthly lifting mechanism, but then tried to debunk them saying that if you used a block and tackle in that manner to lift something as heavy as Ed's coral blocks, so many pulleys would be needed that the friction would make it impossible to move. This logic is exactly like saying that if you downshift your car into first gear, the additional friction would make it impossible to move. Basically, it's the opposite of what's true. Using a block and tackle in this manner is called *mechanical advantage*, and it's what allowed Archimedes to once lift an entire warship full of men using only a block and tackle and his own strength.

Other criticism of the photographs has pointed out that the two obelisks in Coral Castle look like they could be taller than the telephone pole tripod pictured, so it couldn't have been used to raise them to the vertical position. This is just silly. It's like saying you can't lean over, pick up an 8-foot 2×4, and stand it on its end because it's taller than you. But even if the tripods or block and tackles pictured could be proven to be mechanically inadequate for any of the tasks needed to assemble Coral Castle, that doesn't prove magnetic levitation was the only other possible agent. In the nearly 30 years that Ed worked, I'm sure the one rig in the picture was not the only time he ever set up such a system. It is reasonable to allow for the possibility that Ed set up other such rigs at times other than the day these particular photos were taken. There is no reason to conclude that the photographed rig represents his only capability or even his maximum capacity.

An extension of this same argument states that just because Ed used a tripod crane in this picture, doesn't mean that he didn't also use energy vortex powers. This is quite true, of course. It's also true that just because I can dig a hole with a shovel, it doesn't prove that I can't also dig one with *my* energy vortex powers.

Before Ed moved from Latvia at age 26, he grew up in a family of stone masons. Very little is known about what type of stone mason work he did in Latvia, but it probably explains his interest and knowledge of quarrying, cutting, and carving stone. He then lived in Canada and worked as a lumberjack, work which is largely about moving felled trees. A tree the weight of Ed's largest coral block, 30 tons, is not at all uncommon — some trees can weigh hundreds of tons. Before Ed ever started work on Coral Castle, he had a wealth of work experience that gave him all the knowledge he'd ever need to build his creation. It's simply not necessary to invoke made-up mystical powers, aliens, or magnetic vortex energy to explain Coral Castle.

References & Further Reading

Arnold, Dieter. *Building in Egypt: Pharaonic Stone Masonry.* New York: Oxford University Press, 1991.

Editors. "Who's Ed?" *Who's Ed? - Coral Castle Museum.* Coral Castle Museum, 1 Jan. 2008. Web. 16 Jan. 2010. <http://coralcastle.com/whos-ed/>

McClure, R., Heffron, J. *Coral Castle: The Story of Ed Leedskalnin and his American Stonehenge.* Dublin, Ohio: Ternary Publishing, 2009.

Thomson, Peter. "Coral Castle." *Ancient History, Fact or Fiction.* Peter Thomson, 1 Jan. 2005. Web. 24 Jan. 2010. <http://www.peter-thomson.co.uk/coralcastle/coralcastle.html>

Wallington, W.T. "The Forgotten Technology." *W.T. Wallington's Forgotten Technology Official Website.* W.T. Wallington, 20 Jul. 2006. Web. 1 Mar. 2009. <http://www.theforgottentechnology.com/>

30. The Placebo Effect

Got some chronic pain? An itchy rash? Hypertension? Depression? We've got the solution for you. It's guaranteed to have no side effects, but it's an extraordinary treatment for your symptoms. It's the placebo, a completely inert and ineffective intervention, that does nothing to your body at all, except to convince you that it does. Whether it's a pill containing no medication, an expensive looking electronic device that does nothing, an inhaler that provides the same air you're already breathing, or just a manipulative treatment that doesn't manipulate anything, the medical placebo is not only a crucial component of clinical trials, but it can also be an effective medical treatment in itself. After all, placebo is Latin for "I please you."

Most people have heard of the use of placebos in medical testing. When you're testing an actual new drug or device, a control group of subjects receives an identical procedure, but with a placebo instead of the new treatment. If everything is properly controlled, and neither the administrators nor the subjects know which treatment is being used, we get a good understanding of the new drug's actual effectiveness beyond what the placebo provides. But what a lot of people don't know is that even the placebo control group often fares better than groups who receive nothing at all. The fact is that when you *think* you're being treated, you often do experience some level of relief.

There is a very important caveat. As powerful as the placebo effect can be, it does not have therapeutic value in actually treating a physical disease. It also does not usually produce any kind of *measurable* improvement. Its value is almost exclusively for the reduction of symptoms that can be self-reported by the patient. Basically, it can help the way a patient feels, but it can't treat their illness. Exceptions are cases like stress, insomnia, or nausea, where there is an actual condition, but no disease agent is involved.

Placebos are not just pills or drugs. Placebo treatments can also be the application of heat, ultrasound, massage, injection, manipulation, or just about anything else you can do to a body.

Who responds to placebos? Not everyone. In fact, any given placebo treatment is usually only effective on about 1/3 of patients; although nearly everyone will respond to at least one type of placebo. And in fact, most of us do, all the time. Many over-the-counter remedies, and even some prescription drugs, have been found to have effects only comparable to a control placebo. And yet, we perceive that they work, and so we keep going back for more. For example, a review of randomized controlled trials of over-the-counter cough syrups found that no such products work better than a placebo. And so, for all practical purposes, non-prescription cough syrups *are* placebos. That doesn't mean they *don't* work; it means they only work as well as a harmless substance that you *think* will help your cough. And often, that effect is sufficient.

But not all placebos are equal. Some are absolutely more effective than others. It turns out that a wide variety of factors have definite influence over how well a placebo works. If the placebo is a pill, its color, size, shape, taste, and odor all have an impact. In any type of placebo administration, the manner and friendliness of its delivery affect its value. Even when and where the placebo is given can have an effect: An impressive looking office building versus the trunk of a car in a back alley.

- Blue pills are more effective than red pills for calming or tranquilizing. Red pills are more effective than blue pills for stimulation.

- Two pills have more effect than one.

- Pills with a recognized, well-known brand name and packaging are more effective than generic pills. In one large trial published in the British Medical Journal (Branthwaite, A., Cooper, P.), branded aspirin was more effective than generic aspirin, which was more effective than a branded placebo, which was more effective than a generic placebo.

- Expensive treatments are more effective than inexpensive ones.

- The *description* of the placebo's effect is also a powerful factor. Patients who receive a strong warning from the doctor about the strength of the drug have better results than patients who receive a weaker description of the drug's effect. Both groups show better improvement than patients who receive no information about the drug's effect.

- Patients who receive placebos from someone in a white labcoat get better results than when the placebo is administered by someone *not* wearing a white labcoat.

- Better results are obtained from placebos when the doctor spends more time with the patient explaining things.

- The drama and invasiveness of the placebo is a significant factor in its effectiveness. For example, a shot is more effective than a pill. Electric shock is more effective than ultrasound. Acupuncture is more effective than manipulation.

- Paradoxically, placebo treatments that produce unpleasant side effects are more effective than placebos with no side effects.

How do these differences affect placebo controlled trials? If I'm testing a new drug to treat depression, and my drug is a small blue pill in a generic bottle, and the control placebo is a large shiny red pill in a brand name bottle, guess what? There's a good chance that the placebo may actually produce better results than my drug, even though my drug may actually be effective. This can really happen in some cases. The characteristics of the placebo treatment are poorly selected in this example, thus it's not a valid control procedure. Designers of well-performed clinical trials know this, and an important part of their job is to eliminate such variables that can skew the results. The characteristics of the placebo should always match the active treatment as much as possible.

Sometimes, this can include mimicking any side effects that the real drug might produce. There is also a class of placebos called "active placebos". An active placebo is intended to produce the same side effects as the new treatment being tested. An active placebo does contain some active agent, just not one that is intended to treat the patient's condition. Say you're testing a new drug that has the unfortunate side effect of making its subjects dizzy and nauseous. To properly control your trial, you'd need to administer a placebo that makes people dizzy and nauseous also.

Placebos can also work in reverse to produce this same effect. If you tell a patient that an inactive placebo is a powerful drug that will produce negative side effects, it actually can. This is called a nocebo (Latin for "I will harm"). One study of 109 clinical trials in which patients were told the placebo would have adverse effects actually did produce them in 19% of the patients. These side effects were mainly headaches, drowsiness, and weakness: Again, the usual types of conditions that placebos are able to affect.

Since belief is presumed to be the major driver of a placebo's effectiveness at treating self-reportable symptoms, there is some debate over whether the effect also exists in animals. It has been shown that many animals can be calmed by soothing human contact. Dogs and horses in particular will often experience reduced heart rates when they are petted. When animals are taken to certain alternative veterinary practitioners for treatments that we would normally ascribe to the placebo effect in humans, such as veterinary homeopathy or whispering or acupuncture, the treatments are always accompanied by soothing petting by their owners and other kind treatment. Predictably, the animal's mood will typically improve; and just as predictably, the owner will often attribute this to the treatment.

Thus, you might be inclined to argue that the placebo effect isn't needed to explain the animal's improved behavior. And in many cases, you're probably right. There are often overlooked influences present that can explain an animal's improvement, or as in the case of at least one experiment, the animal reacting negatively. In this case experimenters were studying the effect of a drug injected into rats, and found that the rats were more stressed than non-injected rats. Finally employing a placebo injection of saline solution, they determined that the stress was not caused by the drug at all, but rather by the injection procedure.

Since belief in a treatment's effects is the important factor in placebos, it seems unlikely that it could apply to animals, since they lack the understanding of what the treatment is intended to do. Another complication is that many animals can be influenced by learned behavior. For example, if you give rats a mild shock when they eat from the wrong bowl, they can learn to eat from the right bowl. Take the right bowl away, forcing them to eat from the wrong bowl, and rats will exhibit stress because they associate that bowl with the shock. Some have argued that this is an example of the placebo effect, since the rat understands that this food will cause a shock. But since the effect is equally well explained by simple learned behavior and associations, the jury is still out on whether placebos work on animals. What is known is that to effectively test a treatment on animals, you must give the control group a properly designed placebo, just as you would with humans; however this is to eliminate other causes for the treatment's potential effects (like the stress from the scary injection), and it's not intended that the animal will expect to get better the way a person might.

You can actually get real placebo pills if you want them for some reason. In some locales, a pharmacist can sell them to you upon request. In other places you may need a prescription from a doctor, so you'll have to employ your salesmanship skills. If all else fails, you can buy any homeopathic pills, since those are technically placebos. I spoke with one doctor who keeps generic looking vitamin pills on hand, to give his children whenever they insist they're sick and want Dr. Daddy to fix them.

Understanding the power of the placebo effect is crucial to understanding the value of a claimed alternative therapy. If it's well designed and well delivered, an implausible therapy with no clinical value can indeed produce a subjective improvement in the patient's symptoms. To debunk a worthless alternative therapy, it's not necessary to prove that it has no effect at all. Rather, understand that under the right conditions it can, in fact, have a sometimes significant effect; an effect which can almost certainly be fully explainable as a placebo.

References & Further Reading

Bausell, R. Barker. *Snake Oil Science: The Truth about Complementary and Alternative Medicine.* New York: Oxford University Press, Inc., 2007. 23-36, 83-100, 127-166, 207-208.

Chin, Richard, Lee, Bruce Y. *Principles and Practice of Clinical Trial Medicine.* London: Academic Press, 2008.

Drago F, Nicolosi, A., Micale, V., Lo Menzo, G. "Placebo affects the performance of rats in models of depression: is it a good control for behavioral experiments?" *European neuropsychopharmacology : the journal of the European College of Neuropsychopharmacology.* 1 Jun. 2001, Volume 11, Number 3: 209-213.

Rozenzweig, P., Brohier, S., Zipfel, A. "The placebo effect in healthy volunteers: influence of experimental conditions on the adverse events profile during phase I studies." *Clinical Pharmicology and Therapeutics.* 1 Nov. 1993, Volume 54, Number 5: 578-583.

Schroeder, K., Fahey, T. "Should we advise parents to administer over the counter cough medicines for acute cough? Systematic review of randomised controlled trials." *Archives of Disease in Childhood.* 1 Mar. 2002, Volume 86, Number 3: 170-175.

Turner, J., Deyo, R., Loeser, J., Von Korff, M., Fordyce, W. "The Importance of Placebo Effects in Pain Treatment and Research." *JAMA: Journal of the American Medical Association.* 24 May 1994, Volume 271, Number 20: 1609-1614.

31. Attack of the Globsters!

St. Augustine, Florida, 1896 — It came from the sea, its great bulk suspended in the gentle surf, until it began to drag along the sandy bottom. A final push from a larger wave and it took hold upon the beach. Its mass settled and the water receded. The body of *Octopus giganteus,* the largest creature ever to swim in the sea, had washed ashore. Tourists and media flocked to see the five ton remains of the fearsome beast. What became known as the St. Augustine Monster had made its mark on history, and established itself as the first in a long line of creatures collectively called globsters: Great globs of unrecognizable tissue washed up on beaches, too large and shapeless to be anything but sea monsters.

All too often, the default skeptical position has been to brush these off as misidentified whale parts or some other marine life that has decayed to the point of being unrecognizable. And that explanation is certainly true in many cases, probably in the majority of cases. But to be a responsible skeptic, you can't simply ignore the small number of cases that don't fit that explanation. The fact is that a few of these globsters are *not* consistent with the usual suspects, like whale blubber or basking shark carcasses, and that's something skeptics should be aware of. Some of these few globsters that genuinely do defy expert explanation do actually appear to be more consistent with the cryptozoologists' preferred interpretation — that of a legendary, gargantuan, undiscovered cephalopod that they call *Octopus giganteus:* Too big to be a whale part, too shapeless to be from a giant shark. And it's these few of the strangest globsters that are worth a skeptical examination.

There has been a larger number of specimens washed up that have included recognizable anatomy, such as bones and teeth, that have made positive identifications possible, despite the unrecognizable condition of the carcass. Often all that remains are the strongest structures like the spine and fins, sometimes giving the carcass the look of a long, thin sea serpent. When samples of such carcasses have been preserved and later been able to be tested with modern analysis, they've always turned out to be known animals. Some of the frequent culprits include basking sharks, which leave enormous carcasses of incredibly tough cartilage; and beaked whales, which can be huge but have long reptilian-looking snouts that often baffle observers. A few of the best known cases of such beasts have been the Bermuda Blob of 1988, the Japanese Carcass Catch of 1977, Scotland's famous Stronsay Beast of 1808, and the Newfoundland Blob of 2001.

Another important category of globsters is those for which we simply have insufficient data. Many famous globsters have managed to become so without any good evidence: Nobody took a photograph, no samples were preserved, and all we have are verbal descriptions from witnesses. If you do any reading about globsters, one thing you'll learn quickly is that the verbal descriptions vary widely. When the partial carcass of a beaked whale washed up in Santa Cruz, California in 1925, measurements taken by different people placed its length at 20 feet, 35 feet, and 50 feet. Weight is an observation that should be taken with an especially large grain of salt: In all my research I've never once found a case where someone wrangled some enormous scales down onto the beach and actually weighed a globster, yet in virtually every account you'll read claimed weights of 2 tons, 5 tons, even as much as 70 tons in the case of the Suwarrow Island Sea Serpent. When the stories are accompanied by photographs, these weights are often clearly grossly overestimated.

Take all the measurements you read with a grain of salt. They are almost always off-the-cuff guesses by random laypeople, usually passed along second or third hand, and obviously often exaggerated.

But even in the cases where descriptions are good and photographs are available, it's sometimes impossible to make a definite identification without a sample. There are good photos available of many globsters, including the St. Augustine Monster. A local physician, Dr. Dewitt Webb, was the only scientist to see the creature in person, and though his expertise was in medicine and not in ocean creatures, he felt confident enough to identify it as a giant octopus, believing that some of its shredded appendages were the stumps of huge tentacles. Based on this tentative identification, Webb sent photographs to the nation's leading authority on cephalopods, Professor Addison Emory Verrill. At first Verrill thought it might be *Architeuthis,* the Giant Squid. In one article he wrote, Verrill changed his mind and concluded that it could be an octopus; but since it was many times larger than any known species, he assigned to it a new name, and the term *Octopus giganteus* was born.

Interesting, that based on an early assumption by Webb, a man with the best of intentions but the wrong background, only the cephalopod explanation was considered. Should he not also have sent samples and photographs to whale experts and shark experts, and allowed alternate theories? Does the lack of a known explanation force us to conclude that there can *be* no known explanation?

There is a clue to the identity of a lot of globsters hidden within the witness descriptions. Often, you'll read that the creature was covered with white fur, or sometimes brown fur. Sometimes stiff, pointed quills are reported. White fur or quills are an immediate tipoff that what you're looking at it is almost certainly deteriorating connective tissue, which is made of collagen.

Collagen is the most abundant protein in mammals and most other animals. It's a long, sturdy, triple-helix molecule and is the major component of strong connective tissue like bones, tendons, ligaments, and cartilage. Leather is strong because it's mostly collagen. Those indestructible tendons that get in your way when you carve a turkey are a hassle because they're mostly collagen. Many of the ligaments in your body are strong enough to lift a car because they're mostly collagen. Suffice it to say that if you're looking for durable biological material, look no further than collagen.

Collagen is usually the last part of a body to decompose, and marine animals have a lot of it in their skin and in whale blubber. As this collagen-rich soft tissue decomposes, it leaves behind a mat of white fibers, which can have the appearance of dirty, shaggy, white fur. Throw in some tendons and you've got quills too. So keep an eye out for white fur when you read globster accounts; it's a great clue that you're very likely dealing with a big piece of the skin or blubber of a whale or other large animal.

So I found the collagen explanation quite satisfying for a number of the so-called "unidentifiable" globsters. But let's go back to what we were saying earlier, about those few cases where that interpretation truly doesn't fit the photographs. I'm speaking most specifically of the 1896 St. Augustine Monster, which was simply too big, thick, and solid of a mass to be a big piece of blubbery whale skin. I would also consider the 1960 Tasmanian

FAMOUS GLOBSTERS		
Collagen Masses		
St. Augustine Monster	Florida	1896
Tasmanian Globster	Tasmania	1960
Hebrides Blob	Scotland	1990
Four Mile Globster	Tasmania	1997
Whale Blubber		
New Zealand Globster	New Zealand	1965
Tasmanian Globster 2	Tasmania	1970
Bermuda Blob 2	Bermuda	1995
Nantucket Blob	Nantucket	1996
Bermuda Blob 3	Bermuda	1997
Chilean Blob	Chile	2003
Whale Carcasses (various species)		
Suwarrow Island Sea Serpent	Suwarrow	1899
Santa Cruz Sea Monster	California	1925
Tecoluta Sea Monster	Mexico	1969
Newfoundland Blob	Canada	2001
Shark Carcasses (various species)		
Stronsay Beast	Scotland	1808
Cherbourg Carcass	France	1934
Effingham Carcass	Canada	1947
Hendaye Carcass	France	1951
Japanese Carcass Catch	New Zealand	1977
Bermuda Blob	Bermuda	1988
Insufficient Data		
New Harbor Sea Serpent	Maine	1880
Natal Carcass	South Africa	1922
Glacier Island Creature	Alaska	1930
Gourock Carcass	Scotland	1942
Dunk Island Carcass	Australia	1955
Melbourne-Hobart Carcass	Australia	1955
Mann Hill Beach Globster	Massachusetts	1970
Gambian Sea Serpent	Africa	1983
Godthabb Globster	Greenland	1989
North Carolina Globster	North Caro-	1996

Globster, the 1990 Hebrides Blob, and possibly the 1997 Four Mile Globster from Tasmania to fit this category: Well explained as collagen, but of sizes and shapes poorly explained by any *known sources* of collagen. Indeed, in 2004, amino acid analyses of samples taken from the St. Augustine Monster and five other famous globsters confirmed that they were all whale collagen. Probably most of these globsters were chunks of blubber. But is there something else — something bigger and thicker than blubber — that could give the St. Augustine Monster the shape of the body of a giant octopus?

A sperm whale's junk is the lower section of the spermaceti organ in its head. Making up a quarter of a sperm whale's entire mass, the complete spermaceti organ can weigh over ten tons and produces the sperm whale's distinctive head shape. It's encased within a huge, thick, fibrous muscle called the maxillonasalis muscle. Most of the organ consists of a huge sac of spermaceti oil, the stuff most prized by whalers of old. But the bottom third of the organ, weighing up to three tons itself, is the junk. This is a sac with much denser oil, and it was called the junk and discarded because it's filled with extremely tough collagen partitions. Whatever other function the junk might have, sperm whales also use it as a battering ram, having even sunk whaling ships with it, most notably the *Essex* and the *Ann Alexander*. This giant mass of tough collagen fibers is one of the hugest and most durable pieces of anatomy in the entire animal kingdom, and yet it's something almost nobody has ever heard of, or certainly would be able to identify if it washed up on a beach. If you found a shapeless three-ton mass of collagen on a beach, what would *you* think it is? Surely nothing recognizable; it's no wonder they went for the *Octopus giganteus* option.

Might the St. Augustine Monster, or even some of the others, have been sperm whale junk? Of course I can only speculate, but Professor Verrill knew a little better. Dr. Webb sent him a sample, and upon examining it, Verrill said that it was definitely not a cephalod, but rather cetacean tissue. He even said "The whole mass represents the upper part of the head of [a sperm whale], detached from the skull and jaw." And yet, cryptozoology web sites still uncritically embrace the *Octopus giganteus* identification. The takeaway should be that there are still perfectly plausible explanations, no matter how alien the globster seems. When you hear people conclude that there is no natural explanation just because they don't know one, you should always be skeptical.

References & Further Reading

Broad, William J. "Ogre? Octopus? Blobologists Solve an Ancient Mystery." *New York Times.* New York Times, 27 Jul. 2004. Web. 25 Jan. 2010.
<http://www.nytimes.com/2004/07/27/science/27blob.html?ei=5090&en=db3de26bb5c38e21&ex=1248667200&pagewanted=1>

Carr, S., Marshall, H., Johnstone, K., Pynn, L., Stenson, G. "How To Tell a Sea Monster: Molecular Discrimination of Large Marine Animals of the North Atlantic." *The Biological Bulletin.* 1 Feb. 2002, Volume 202 Number 1: 1-5.

Carrier, David. "The face that sank the Essex: Potential function of the spermaceti organ in aggression." *The Journal of Experimental Biology.* 5 Apr. 2002, 205: 1755–1763.

Pierce, S., Smith, G., Maugel, T., Clark, E. "On the Giant Octopus (Octopus gigmteus) and the Bermuda Blob: Homage to A. E. Verrill." *The Biological Bulletin.* 1 Apr. 1995, Volume 188 Number 2: 219-230.

Verrill, A. E. "The Florida Sea-Monster." *The American Naturalist.* 31 Dec. 1897, Volume 31, Number 364: 304-307.

32. Was Chuck Yeager the First to Break the Sound Barrier?

We all know the story of how Captain Chuck Yeager opened the throttles of the Bell X-1 *Glamorous Glennis* in October, 1947. Breaking the sound barrier was to aviation what Neil Armstrong's first step was to the space program: No matter how many others went higher or faster later, it will always be that seminal, unassailable "first" that can never be topped. Yeager's name will always sit atop every list of record-breaking pilots, up there by himself in his own special stratosphere. But: Was he really the first pilot to fly faster than sound?

Plenty of stories out there say Yeager wasn't the first. How do we know what to believe? Do we accept the popular official story, or do we give credibility to the other claimants with good evidence of their own? Today we're going to point our skeptical eye at some of these other claims, and see who really deserves the credit.

There are certainly many pilots who approached the sound barrier but didn't live to tell about it. The years preceding Yeager's flight were among the most exciting in aviation history, as World War II drove aeronautic advancement like never before. Planes that had been shot down often entered the transonic realm as they plummeted, and were torn apart by the resulting shockwaves. Dive bombers had to have special air brakes developed to prevent them from breaking up, which sometimes happened anyway. Of the many pilots who toyed with the sound barrier in WWII — all unintentionally, of course — most never survived the adventure.

During WWII, engineers didn't yet have any flight test experience that taught us how to design aircraft capable of supersonic speed. Even in 1947, Yeager's X-1 was designed after a 50 caliber bullet, known to be stable at supersonic speeds. WWII had seen widespread use of the German V-2 rockets, which were supersonic, so we knew such flight was possible. But the V-2 was ballistic, it didn't require a controllable airframe; and designing a supersonic controllable airframe was the problem for aeronautical engineers. The main issue is called shock stall, and it's what happens when a control surface approaches the speed of sound. A shockwave forms around the control surface, rendering it useless, and the pilot has no way to control the aircraft.

Propeller aircraft can never reach the sound barrier, since the tips of propeller blades hit the sound barrier before the rest of the plane does. The propeller blades go into shock stall, and the plane can no longer accelerate. There are many claims of propeller driven dive bombers breaking the sound barrier during WWII, but these have to all be considered implausible. Approaching the sound barrier, an airplane is already well above its terminal velocity, the speed at which drag matches the acceleration imparted by gravity. Propellers are shock stalled, and there is neither thrust nor gravity available to accelerate a diving airplane past a certain point. As any aircraft approaches the speed of sound, airflow over some parts of the plane will exceed Mach 1 and create shockwaves. These shockwaves cause intense buffeting. Many propeller driven WWII fighter planes, including the Supermarine Spitfire, the Lockheed P-38 Lightning, and the North American P-51 Mustang, experienced these effects at Mach 0.85. Similarly, jet engines of the day were not designed to work with supersonic airflow entering through the compressor vanes; such engines would flame out.

However, one particular fighter plane of WWII was not driven by either propellers or jets: The German rocket powered Messerschmitt 163 Komet. The Komet was designed by the great Alexander Lippisch, a pioneer of delta wings and ramjets. By the end of WWII, Lippisch had a test glider of a supersonic ramjet powered aircraft actually undergoing flight tests. He understood the requirements of supersonic flight. The Komet was designed to fly as fast as possible while staying under the critical limits at which trouble happens. The Komet's delta wing was exceptionally thin. This delays the onset of shock stall over the primary airfoil, and allowed the Komet to remain stable up to Mach .85. However, Lippisch had no answer for shock stall at the control surfaces on the trailing edge of the delta wing, and so the Komet was destined to remain subsonic. In combination with the relatively low 3,800-lb thrust of the Komet's Walter rocket engine (about half the power of Yeager's X-1), the Komet had little expectation of going supersonic, except in an uncontrollable, powered dive which would probably be unrecoverable.

Nevertheless, stories persist of Komet pilots breaking the sound barrier, years before Yeager did. Komet test pilot Heini Dittmar, flying an early prototype in 1941, reached an officially measured speed of 1,004 kph in level flight. This was probably around Mach .95 but we don't know for sure since the flight was classified until after the war and the altitude is unknown. However, Dittmar made the flight at partial throttle to avoid buffeting, using an engine only half as powerful as that which went into production. But just because later versions were more powerful doesn't mean they wouldn't run into exactly the same limitations at the same top speed. One unofficial report claims that Dittmar hit 1,130 kph in 1944 (Mach 1.06), and another states that in a steep dive he created sonic booms that were heard on the ground. However, these stories first appeared in 1990 book written by Dittmar's friend Mano Ziegler, and do not have contemporary corroboration or documentation. But the best evidence against Komets breaking the sound barrier is the fact that the Allies did capture all of the program's classified data, and no supersonic flights were ever recorded, even in secret.

Claims you'll find on the Internet that "Komet pilots routinely broke the sound barrier" cannot be given much weight, given the aircraft's limitations well understood by Alexander

Lippisch. As an aircraft approaches the speed of sound, the shockwave over the wing moves the center of lift backwards and the plane noses down. This is a condition called Mach tuck. Normally you'd pull back on the stick to correct this, but conventional elevator controls on the trailing edge of the tailplane would be unable to get any bite, since the elevators would be shock stalled. The only way out of Mach tuck is to use an all-moving tailplane to trim back to level. With its delta wing, the Komet had no tailplane at all, let alone an all-moving tailplane. It had fabric-covered elevons on the trailing edge of the delta wing, which would always be shock stalled. Even if a pilot opened his Komet's rocket engine to full throttle to muscle his way past the sound barrier, Mach tuck would send him tumbling out of control irrecoverably, and probably destroy the airframe.

It's also important to be aware of a limitation of early airspeed indicators. One built for subsonic speed is probably going to give unreliable readings in the presence of shockwaves. A phenomenon called compressibility error gives inaccurately high airspeed readings as the aircraft approaches the speed of sound. This error is called Mach jump. To counter this, supersonic aircraft use a Mach indicator instead of an airspeed indicator. The speed of sound at any given pressure and altitude is determined primarily by temperature. A Mach indicator is essentially an airspeed indicator mounted on an aneroid diaphragm to correct for static air pressure. Since Mach and airspeed are both dependent on temperature, they cancel each other out and no temperature diaphragm is needed. Komets had airspeed indicators, not Mach indicators; and so even the speeds logged by the German test pilots are probably incorrectly high.

One of the best known claims to the sound barrier comes from German WWII fighter pilot Hans Guido Mutke, flying perhaps the most devastating fighter of the war, the Messerschmitt Me-262. The 262 was the first true operational jet powered fighter plane in the world, sporting twin BMW 003 turbojet engines mounted below the swept wings. Although the 262 entered the war too late to have any real impact, it boasted a 5:1 kill ratio against allied fighters. Mutke was cruising at 36,000 feet when he began a steep dive under full power. With his airspeed indicator pegged at its limit of 1,100 kph (just over the speed of sound, but remember the airspeed indicator problem), Mutke reported severe buffeting and loss of control. Suddenly the buffeting stopped and he regained control, with the airspeed indicator still pegged; and it's this that could indicate he had broken the sound barrier. Unfortunately his engines flamed out, not being designed for supersonic speeds, and he slowed, and the severe buffeting returned. Finally his speed dropped enough that he regained control again and was able to restart his engines. He returned to base, and it was found that his aircraft had lost many rivets, and its wings had become so distorted that the plane had to be scrapped.

Mutke never understood what had happened until Chuck Yeager's flight was declassified and the supersonic flight profile became known: Severe buffeting while approaching Mach 1, then the shaking stops above Mach 1, and then resumes upon deceleration below Mach 1. But unfortunately for Mutke, there was not, and could not have been, any independent verification of his speed or of the period of smooth supersonic flight. Nobody denies the damage done to his plane during the buffeting period, but supersonic flight was not necessary for this to happen.

The designer of the Me-262, Willy Messerschmitt, always stated emphatically that the 262 was incapable of supersonic flight. In flight tests, he found that at Mach 0.86, the 262 experienced Mach tuck: It lost control and assumed a nose-down attitude that could not be corrected by the pilot, and throttling down was the only way to resume control. The 262 only had conventional elevators on the trailing edge of its tailplane, like all aircraft of the day, so these would have shock stalled and not been able to correct the Mach tuck. But the 262 also had an additional feature: The tailplane was actually all-moving for trim purposes. This was a separate electrically operated control, and it was normally used to keep the plane level as its fuel supply was consumed. Mutke reported that he actually employed this all-moving trim control in order to get out of the nose-down state, a technique which may not have been considered in Messerschmitt's own tests. Mutke's report was given additional credibility in 1999, when computer modeling and scale model wind tunnel testing conducted at Munich Technical University found that the 262 was capable of reaching, and passing, Mach 1.

So, while we cannot prove or disprove Mutke's claim, it is possible that he did reach supersonic flight. However, like in sports, it's not what happened, it's what the referee *says* happened that matters. Mutke's feat was unverified and unofficial, and certainly unintentional; so even if he did break the sound barrier before Yeager, it "doesn't count."

There are two other flights that don't count, both accomplished by George Welch, a civilian test pilot for North American Aviation. On October 1, 1947, just 13 days before Yeager broke the sound barrier in the X-1, Welch took the new XP-86 fighter prototype up for its maiden flight. In a powered dive from 35,000 feet, Welch reported Mach jump on his airspeed indicator, showing that he was traveling supersonic. Anecdotal stories say that a sonic boom was heard on the ground. Welch's airspeed was not being officially recorded, and no official record states that he broke the sound barrier. If he did, it was either unverified or classified. Welch believed that he did, and to hammer the point home, he gave a repeat performance. While Yeager was strapped into the X-1 still attached to its B-50 mother ship, just before the historic flight, Welch again put his XP-86 into a steep dive. Some stories say that he buzzed the B-50 close enough for those onboard, Yeager included, to hear his sonic boom. He made a 4g pullout from his dive, and those same stories say that his sonic boom was louder than Yeager's just 20 minutes later.

There is no engineering reason to doubt Welch's claim. The history books credit George Welch with breaking the sound barrier in the XP-86 in a dive six months later on April 26, 1948, with official measurements and a proper Mach indicator on board. Did he do the same thing before Yeager's flight? He may well have, and a lot of people say he did. But there's a significant difference between Yeager's flight and those of Welch, Mutke, Dittmar, and probably others. Their claims to the sound barrier were all in dives, and were transient at best. *Glamorous Glennis*, on the other hand, was the first aircraft capable of sustained supersonic level flight. Sure, being the first to break the sound barrier becomes less glitzy when you have to pile on qualifications. But every aviation milestone has been an incremental one; few are truly revolutionary. Yeager, Welch, Mutke, and Dittmar all made real contributions to the science of aviation. All had "the right stuff". At some point, any lines you draw to separate their achievements come down to semantics; yet you still have to draw those lines somewhere. Flights have to be official, they have to be verifiable, and they should demonstrate a deliberate capability. And so, while it's a virtual certainty that the sound barrier was broken by someone somewhere in some circumstance, Chuck Yeager's flight of the Bell X-1 is the only flight to meet all the criteria of a true aviation first.

REFERENCES & FURTHER READING

Hilton, William F. *High-Speed Aerodynamics*. London: Longmans, Green and Co., 1952. 15-19, and Ch. 6.

L.J. Clancey. *Aerodynamics*. Ney York: Wiley, John & Sons, Incorporated, 1975. 283, 415.

Mason, W.H. *Configuration Aerodynamics*. Blacksburg, VA: Virginia Tech, 2006. Chapter 7.

Rotundo, Louis C. *Into the Unknown: The X-1 Story*. Washington, D.C.: Smithsonian Institute Press, 1994.

Wagner, Ray. *The North American Sabre*. London: Macdonald, 1963.

Yeager, Chuck, et. al. *The Quest for Mach One: A First Person Account of Breaking the Sound Barrier.* New York: Penguin Studio, 1997.

33. NLP: Neuro-linguistic Programming

Today we're going to point our skeptical eye at Neurolinguistic Programming, a New Age communication technique intended to facilitate the exertion of influence. Is it science, or is it another spin-the-wheel-and-invent-a-new-self-help-system disguising its marketing within scientific sounding language?

It was the early 1970's, and a young psychology student at the University of California, Santa Cruz was spending another late night in the lab. Richard Bandler's assignment was to transcribe hours and hours of psychotherapy sessions from the maverick German psychiatrist Fritz Perls. After transcribing until his hands were about to fall off, Bandler noticed an interesting pattern in the way Perls spoke to his patients. Perls had an odd — almost annoying — habit of taking his patients' comments and going back over them with very specific questions, forcing the patients to closely re-examine their wording. Sometimes it seemed that you couldn't make the simplest remark without Perls raking you over the coals. What made you choose this word; what are the implications of your statement? Perls would force his patients to confront the causes and motivations of even the most casual remark. Bandler noticed that this technique had a dramatic effect. Patients would eventually be ground down to the point that they were unable to explain themselves, leaving something of an internal void, and became exceptionally receptive to Perls' suggestions to fill that void. Rather than resenting what might be called harsh cross examination, patients instead tended to embrace the process; and Bandler found that taken as a whole, Perls' technique seemed highly effective.

Bandler reported his discovery to John Grinder, who was a linguist at Santa Cruz. Grinder was intrigued. The two discussed Bandler's findings at length, and decided to look for other incidences of the same pattern. They found them in the psychotherapy sessions of pioneering family therapist Virginia Satir. Believing that they'd stumbled onto something significant, Bandler and Grinder documented and codified the technique, and named it the Meta Model. Built largely around the Meta Model, the two men published the first two of many books to come in 1975. They heralded their discovery as a breakthrough in psychotherapy that would "help people have better, fuller and richer lives." (Keep in mind that this alleged breakthrough in psychotherapy was created by an undergrad and a linguist, neither of whom was a psychotherapist; though Bandler did go on to get an MA in psychology.)

They then built upon their Meta Model with a very different communication technique that they learned by studying the work of hypnotherapist Milton Erickson. Erickson's style was the polar opposite of the high pressure of the Meta Model. What he did was to give general suggestions to his hypnotherapy clients. He wouldn't give specific directions like "You feel happy," instead he'd give a suggestion like "You're free to feel this way if you want to." Not "Put the cup on the counter," but "Consider other places you might like to put the cup, somewhere over there for example." In this way, Erickson was able to guide the client through to his desired destination, but by leaving all the specific steps to get there up to the client, thus empowering them. Bandler and Grinder called this the Milton Model. They found both to be effective tools for influencing others.

Together, the Meta Model and the Milton Model formed the basis for what they came to call Neuro-linguistic Programming. Bandler and Grinder were up to five books by the time they published their Milton Model, and from then on, their subsequent books covered their whole umbrella of Neuro-linguistic Programming, shortened to NLP. By now, the books were being published by Bandler's own publishing company, Meta Publications. They also offered training workshops and classes, marketed at first through psychology trade publications. But it turned out their business came not from the industry, but from business managers, sales professionals, and New Age enthusiasts. NLP grew from the same roots, and shared many of the same customers, with EST and Esalen, also located in the same region around the northern California coast. Throughout the 1970's, such groups peddled self-help philosophies typically ignored by the mainstream. Bandler, Grinder, and the group of associates that grew around them became wealthy and successful, until the early 1980's when trademark disputes, mutual lawsuits, and Bandler's trial for the cocaine-fueled murder of a prostitute (for which he was acquitted) caused all the NLP leaders to splinter off from one another. Today the term NLP is in the public domain, and most of the original founders still publish their own material and teach their own classes using the term, but there is no one organization that owns the trademark.

I've read a fair amount about NLP, and my analysis of the Meta Model is pretty simple. I'd describe it as a confrontational manner of speaking intended to dominate a conversation by nitpicking the other's persons sentences apart. For example, if it's a good day and all is well, I might be inclined to make an offhand, general comment like "I feel pretty good today." The Meta Model response to that is "What specifically makes you feel good?" And, I don't really know. I don't really have a single, specific answer. And whatever I do come up with gets attacked the same way: "Exactly why does that make you feel good?" And suddenly I'm on the defensive; I'm being made to feel that I'm in error, the position I've taken is revealed to be unsupported; and I'm now putty in the NLP guy's hands. Basically, it's being a condescending jerk in the way you talk to someone, in order to exert influence. That's the Meta Model. It's not psychotherapy; it's high-pressure sales. The Milton Model takes a different road to the same destination: low-pressure sales.

And it's not just sales. It's negotiation in business. It's gaining the upper hand in interpersonal relationships. It's being an effective manager or sports coach. But — and this is the big "but" — despite the claims of those who sell NLP books and seminars, it is *not* part of modern psychotherapy. Russia and the UK do have professional associations of NLP practitioners, but these are composed largely of people selling books and seminars, and only rarely of credentialed psychiatrists. In 2005, the *Australian and New Zealand Journal of Psychiatry* published the results of a comprehensive study of all the publications regarding NLP and similar modalities, which it grouped together under the term "power therapies". The article states:

> *Advocates of new therapies frequently make bold claims regarding therapeutic effectiveness, particularly in response to disorders which have been traditionally treatment-refractory. This paper reviews a collection of new therapies collectively self-termed 'The Power Therapies', outlining their proposed procedures and the evidence for and against their use. These therapies are then put to the test for pseudoscientific practice... It is concluded that these new therapies have offered no new scientifically valid theories of action, show only non-specific efficacy, show no evidence that they offer substantive improvements to extant psychiatric care, yet display many characteristics consistent with pseudoscience.*

It seems the only mentions of NLP to be found in mainstream journals are critical, when they can be found at all, outside of the hypnotism and other fringe journals. Even way back in 1987, the *Journal of Counseling Psychology* published an article that:

> *Examines the experimental literature on neurolinguistic programming (NLP). [The authors] concluded that the effectiveness of this therapy was yet to be demonstrated. Presents data from seven recent studies that further question the basic tenets of NLP and their application in counseling situations.*

Dig far enough and you can find publications that support the therapeutic use of NLP, albeit from journals of varying repute. Wikipedia's article on NLP provides a long list of such articles, so if you wanted to state the case that NLP is science, it would be easy to go there and back yourself up. Well, of course, Joe Blow on the street has no real way of knowing which side he should believe, so this is one case where I'd recommend looking at the meta analyses: Studies that attempt to summarize all the articles out there. The largest of these (that I could find) was done by Michael Heap in 1988:

> *If the assertions made by proponents of NLP about representational systems and their behavioural manifestations are correct, then its founders have made remarkable discoveries about the human mind and brain, which would have important implications for human psychology, particularly cognitive science and neuropsychology. Yet there is no mention of them in learned textbooks or journals devoted to these disciplines. Neither is this material taught in psychology courses at the pre-degree and degree level.*

Heap also found that when he asked colleagues about NLP, they generally hadn't even heard of it. Whatever else you want to say about NLP, the fact is that it is not part of mainstream psychology. That doesn't make it wrong or useless; it just means that it's not part of established, practiced science.

So really, what we have with NLP boils down to just another pop-culture, New Age, self-help system that disingenuously markets itself as science. Read this book and you'll be a better manager, a better salesman, more successful. The promise of results — be they money, success, interpersonal, psychological — is a red flag that you're solidly outside the world of professional psychology, or any other branch of medical science. If any doctor or other profession ever guarantees you results, or tells you that goals are only a few simple steps away, you have very good cause to be skeptical.

References & Further Reading

Carroll, R. The Skeptic's Dictionary: A Collection of Strange Beliefs, Amusing Deceptions, and Dangerous Delusions. Hoboken, NJ: John Wiley & Sons, 2003. 252-257.

Devilly, G. "Power Therapies and possible threats to the science of psychology and psychiatry." Australian and New Zealand Journal of Psychiatry. 7 Jun. 2005, Volume 39, Issue 6: 437-445.

Editors. "List of studies on Neuro-linguistic programming." Wikipedia. Wikimedia Foundation, Inc., 8 Jun. 2006. Web. 26 May. 2009. <http://en.wikipedia.org/wiki/List_of_studies_on_Neuro-linguistic_programming#Generally_supportive>

Heap, M. Hypnosis: Current Clinical, Experimental and Forensic Practices. London: Croom Helm, 1988. 268-280.

Morgan, D. "A Scientific Assessment of NLP." Journal of the National Council for Psychotherapy & Hypnotherapy Register. 14 Mar. 1993, Volume 4, Number 2: 3-4.

Platt, G. "NLP - Neuro Linguistic Programming or No Longer Plausible?" Training Journal. 12 May 2001, Volume 1, Number 23: 10-15.

Sharpley, C. "Research Findings on Neurolinguistic Programming: Nonsupportive Data or an Untestable Theory?" Journal of Counseling Psychology. 7 Jan. 1987, Volume 34, Number 1: 103-107.

34. The World According to Conservapedia

Today we're going to look at a surprising relative newcomer on the World Wide Web: Conservapedia, a Young Earth Christian encyclopedia, built on the model of Wikipedia. But I caution you: We need to start by setting the stage with the appropriate tone. Let's browse to Conservapedia's article on the theory of evolution. What picture do you think they might have on this page? An image of a DNA helix? A picture of Charles Darwin? Wrong: It's a photograph of Adolf Hitler screaming into a microphone. No fooling. *(Conservapedia has since edited that page, putting Darwin's picture at the top where it belongs, and moving Hitler way down toward the bottom of page, where he still doesn't belong.)*

Conservapedia's purpose is to be a conservative Christian version of Wikipedia that promotes Young Earth anti-science. But in an effort to distance themselves from this egregious intellectual dishonesty, they've posted a special page called *How Conservapedia Differs from Wikipedia* that lists an entirely *different* set of reasons for its existence, that makes no mention of any conservative Christian agenda, as if such a thing has never occurred to them. The page is mostly a list of false charges leveled against Wikipedia. They say Wikipedia is a money-making scheme; Wikipedia encourages long-winded, obfuscating articles; it contains gossip, journalists' "biased" opinions, obscenities and pornography; Wikipedia subjects its articles to extreme liberal censorship; the purpose of Wikipedia's talk pages are to create a forum for its editors to bully its users; and my personal favorite, that Wikipedia monitors the personal blogs of its readers and bans them from using Wikipedia if they exercise their free speech on their blogs!! Well, if nothing else, these charges certainly do distract attention from Conservapedia's Christian anti-science bias.

Another difference between the two is their size, both in number of articles and length. For example, on the date I checked, Google listed 50,000 pages on Conservapedia.com, and 14,200,000 on Wikipedia.org, giving Wikipedia almost 300 times the number of pages as Conservapedia. To get a fair estimate of the difference in comprehensiveness of their articles, I did a number of random searches. Wikipedia's article on Walt Disney is 8,570 words long. Conservapedia's article is only 195 words long, and contains little more than an anecdote about his wedding, a list of 8 of his movies, and a warning that since his death, The Walt Disney Company has promoted pro-gay activism. This was just one random search off the top of my head, and I was struck by Conservapedia's ability to siphon all the juice out of any topic until only the bitterest right-wing dregs remain.

But Conservapedia is not really about finding liberal skeletons in closets; it's really about promoting Young Earth fundamentalism. For example, their article on the planet Earth contains no fewer than 9 Bible references, and almost all of the external references are from (not scientific sources as you'd expect) but other Young Earth fundamentalist web sites, mainly AnswersInGenesis.org, CreationScience.com, and CreationOnTheWeb.com. The age of the Earth is given with a pretense of also presenting real science as follows:

> *Young Earth creationists believe, on the basis of the biblical account in Genesis and biblical geochronologies, that the entire Earth, including animal, plant, and human life, was formed in six days, around 4000 B.C. Mainstream scientific journals, committed to a naturalistic worldview, contend this*

view. Most scientists believe that the Earth formed by natural processes instead of being supernaturally created. However, as one scientist noted, "... most every prediction by theorists about planetary formation has been wrong."

Note the assertion that the mainstream scientific journals "contend" the Young Earth story because they're "committed" to a worldview that is not about promoting Christianity. I commend Conservapedia for conceding that most scientists are not of the opinion that the Earth was formed supernaturally, but even then they throw in a quote from CreationScience.com that most scientific theories about the Earth are eventually proven wrong, and they bolster this quote with an argument from authority by crediting it to "one scientist".

Conservapedia has a fine article on radiometric dating. It says almost nothing about radiometric dating; it lists only five types, giving no useful information about what they are, what they're used for, or how and why they work; the entire article is simply a lecture that they are all uselessly unreliable. This argument is made with explanations like the following:

This formula depends on the laws of physics remaining constant over time... Some creationists have argued that God increased the rate of potassium-argon decay during the first few days of Creation, thus causing the potassium-argon dating method to give erroneously old date readings.

They didn't give a Biblical citation for that particular factoid. I don't recall exactly which chapter and verse discussed God's manipulation of potassium-argon decay rates. If we're playing Name the Logical Fallacy, this is called a special pleading. A special pleading states that my claim is true because some higher power (that you can't comprehend) *makes* it true, no matter what you say. Potassium-argon dating is unreliable because God will do whatever is necessary to make your readings wrong. *Really.*

They have a page of rules for editors, but predictably, they call it the Commandments. Most of them are good, what you'd expect from a wiki; like always cite your sources (though almost all sources cited in Conservapedia articles are from Young Earth Creationist web sites), your posts should be clean, concise, no foul language, no personal opinions or advertisements. But I thought this was funny:

> When referencing dates based on the approximate birth of Jesus, give appropriate credit for the basis of the date (B.C. or A.D.). "BCE" and "CE" are unacceptable substitutes because they deny the historical basis.

It's actually more than just funny. CE and BCE, similar to the better known BC and AD, stand for the "common era" and "before the common era" of the proleptic Gregorian calendar. It is the international date standard ISO 8601. BCE and CE have been in use since the year 1615, beginning with non-Christian cultures, and has since become the international standard because it does not force everyone to comply with one particular religious standard. When he read a student's paper that used common era notation, Christian activist Andrew Schlafly founded Conservapedia in 2006, initially as an online resource for homeschooled Christian students.

Schlafly felt that teaching evidence-based science represents a biased view that wrongly excludes Christianity, and so his vision of Conservapedia was to avoid this bias by looking at every topic through a conservative Christian perspective. And the more you go through Conservapedia, the more you see that avoiding bias is what it's really about; but a very curious and specific brand of avoiding bias. It avoids the bias of giving fact more validity than fiction. Throughout its articles, Conservapedia presents both the Young Earth version of a subject, and then the scientific version, and treats them more or less equally, as if fallacy is no less valid than fact. To find another example, I looked up Tyrannosaurus Rex to see what Conservapedia has to say on when it lived:

> Young Earth Creationists believe that they became extinct sometime since the Great Flood, dated to approximately 4,500 years ago. Evolutionary scientists believe that the T-rex lived at the end of the Cretaceous period, dated to ap-

proximately 65 million years ago, and that modern birds are the descendants of dinosaurs such as T-rex.

You might as well say "One small extremist fringe group believes X, but science tells us Y." Conservapedia presents these as equally weighted, competing theories which should both be taught to students under the guise of "presenting both sides of an issue". Science does not have two sides. Science has one side. If Conservapedia truly wants to avoid bias, then why are they only presenting the Young Earth fundamentalist belief alongside science? Why not also present the Islamic story, the Mandinka story from Africa, the Voodoo story, the Hawai'ian story of Pele, or Scientology or Raëlism? I'll tell you why. It's because Conservapedia exists only to promote Christian Young Earth anti-science.

Why would anyone want to teach anti-science? The vast majority of Christians worldwide accept the scientifically derived age of the Earth and the universe, and accept modern sciences, and many work in scientific fields. What do most Christians think of Conservapedia? Do they feel it gives them a bad name? Are they primarily supportive of its proselytizing mission, and less concerned with the accuracy of whatever facts it presents?

My sense is that most Christians are not really aware of Conservapedia. Its take on science certainly does not represent that of most Christians, let alone most conservatives, since we know that the Young Earth crowd is only a small fringe minority. The danger is that Conservapedia's mere presence and self-assertion of unbiased authority might mislead an uninformed web surfer into accepting that some of these Young Earth myths are valid science, or even just give the impression that the Young Earth thing is more widely accepted than it really is. Both of these are crimes against intelligence, and anyone who contributes to Conservapedia is guilty of willfully eroding the collective intellect, to the detriment of all.

References & Further Reading

Daniels, J. *Cyber Racism: White Supremacy Online and the New Attack on Civil Rights.* Lanham, Maryland: Rowman & Littlefield Publishing Group, 2009. 132.

Editors. "Evolution." *Conservapedia.* Conservapedia, 15 Sep. 2008. Web. 15 Sep. 2008. <http://www.conservapedia.com/Evolution>

Irvin, D., Sunquist, S. *History of the World Christian Movement.* New York: Continuum International Publishing Group, 2001. p. xi.

Johnson, B. "Conservapedia - the US religious right's answer to Wikipedia." *The Guardian.* 2 Mar. 2007, International section: p18.

Simon, S. "A conservative's answer to Wikipedia." *Los Angeles Times.* Los Angeles Times, 19 Jun. 2007. Web. 19 Jun. 2007. <http://articles.latimes.com/2007/jun/19/nation/na-schlafly19>

35. High Fructose Corn Syrup: Toxic or Tame?

Today we're going to pop open a can of root beer, pour it over some ice cream, and enjoy a nice float from these two of many foods sweetened with high fructose corn syrup, or HFCS, and see what happens. Although decades of experience tell us that nothing much will happen that wouldn't happen with any other equally calorific food, a vocal and growing minority charges that HFCS is a dangerous chemical poison. It causes obesity and diabetes, and for just about any disease you can find on the Internet, someone claims that HFCS is the cause. There are even conspiracy theories that Big Corn gets kickbacks from Big Pharma to keep people sick. So what's the truth? What are the real differences between sugar and high fructose corn syrup? What are its pros, and what are its real cons?

For some reason, high fructose corn syrup has become a huge public worry. It's been one of the most common questions I get emailed:

Do I need to be concerned? Can you point your skeptical eye at the evil food producers who are poisoning us with HFCS?

Search the Internet and you'll find all kinds of calls for boycotts and 30-day HFCS-free diets. Even the renowned expert on health and nutrition, filmmaker Michael Moore, has warned that the President should make the elimination of HFCS his number three priority. (I know that whenever I seek advice on healthy living, I go to someone who looks and eats like Michael Moore.)

Now I don't want to bore anyone to death with the whole chemistry thing, but here's a ten-cent definition of terms. Carbohydrates come in basic molecules called monosaccharides, or single sugars. The two monosaccharides we're discussing are glucose and fructose. Regular table sugar is a disaccharide of glucose and fructose, which means that the two monosaccharides are chemically bound into a larger, more complex disaccharide molecule called sucrose. That's sugar. HFCS consists of the same two monosaccharides, only they're just mixed in together, the molecules are not bound. This means that HFCS can come in different blends. The more fructose relative to glucose, the sweeter it is. HFCS 55, which is 55% fructose, has a sweetness comparable to sugar and is used mainly in soft drinks. HFCS 42 is 42% fructose, and is a little less sweet than sugar and is used in most other foods.

When you consume regular sugar, sucrose, the first thing your digestive system does is break the chemical bond and separate it into glucose and fructose. So once saccharides are in your body, it makes very little difference whether they came in as table sugar or as HFCS. You can also cook table sugar, and unbind the saccharides that way. The corn lobby is always saying that HFCS is nutritionally the same as sugar, and this is what they're talking about. The chemistry is actually quite simple. So why the controversy, and why all the scaremongering about the terrors of HFCS?

The fact is that there *is* huge correlation between HFCS consumption and obesity, and all sorts of obesity related conditions like diabetes and heart disease. Nobody disputes that. The problem arises when people make the common error of mistaking correlation for causation. There's an equally valid correlation between obesity and dirty dishes. The *cause* of obesity and obesity related diabetes is overeating more calories than you burn. It makes no difference whether you overeat food containing pure cane sugar, food containing HFCS, or organic spinach: Too many calories is too many calories, and you'll become obese and suffer the same obesity related complications no matter what you ate to get you there. Fat is fat.

So if it doesn't matter, why do American companies put HFCS in so many food products? The answer is simple: Farming conditions here are generally better for corn than for sugar. We have to import a lot of our sugar, mainly from Brazil, Mexico, and the Dominican Republic. To protect American corn farmers, we hit those sugar imports with tariffs. In retaliation, those countries put similar tariffs on the HFCS they import from us. Presto, HFCS stays cheap in the US, and sugar stays cheap in Latin America. Thus, Mexican Coke is made with sugar, and American Coke is made with HFCS.

High fructose corn syrup is not only cheaper for American companies, its being a liquid makes it a lot handier to use. It's easier and cheaper to transport, for one thing. It has certain advantages in baking, browning, and fermentability. It doesn't recrystalize after baking like sugar can, and makes foods moister. OK, fine, so what are the disadvantages of HFCS?

Real disadvantages are pretty humdrum, obvious things like it's not as handy to drop a spoonful into your coffee or to keep in the little paper packets on a restaurant table. I found a huge number of bogus claims about it. Usually they misunderstand or misrepresent the chemistry, or they focus on the correlation instead of the cause. Here's one that gets the chemistry wrong:

> *Drinking high-fructose corn syrup ... increases your triglyceride levels and your LDL cholesterol. These effects only occurred in the study participants who drank fructose -- not glucose.*

Regular corn syrup is all glucose. High fructose corn syrup is basically half glucose and half fructose, exactly the same as table sugar. Neither of these is pure fructose, which seems to be what these study participants drank (no source was given). It's a common misconception that because of its name, high fructose corn syrup is composed largely or entirely of fructose. It's not.

Dr. Oz, the thoracic surgeon whom Oprah promotes as a health expert on everything, also gets the chemistry wrong. He says:

> *...The body processes high-fructose corn syrup differently than it does old-fashioned cane or beet sugar, which in turn alters*

your body's natural ability to regulate appetite. It blocks the ability of a chemical called leptin, which is the way your fat tells your brain it's there.

This is an effect of fructose. Since HFCS and cane sugar have the same amount of fructose, they have the same effect on your leptin, which is a hormone that helps to regulate your metabolism. This particular charge is the one that's closest to true. Since the fructose in sugar is still bound to the glucose when you eat it, it does take some time before your body breaks the bond and the fructose is freed — anywhere from a few seconds to perhaps an hour. It's therefore plausible that there could be some delay in feeling full when you eat food sweetened with HFCS relative to if you'd eaten an equivalent quantity of sugar, but all the studies have thus far shown no difference.

There's one thing that complicates this type of research, and that's that the corn industry, including major producers of HFCS products like soft drinks, often fund some of the research that finds no special risks associated with HFCS relative to other sweeteners. My own experience reading research is that it's rarely clear who funded it. The exception is that when a reputable journal publishes such an article, where there would be a clear bias or conflict of interest, it's pointed out up front to remove any possibility of impropriety. Outside of this, I generally have no idea who funded a given study. You can go to the authors' web sites and find out who they work for, and sometimes you can get an idea. But should you always assume that scientists are not actually doing research, but merely parroting the commercial desires of whoever funds their grant? No doubt a small number do, but these are rarely or never the ones whose work passes the scrutiny of peer review and makes it into a reputable journal. They are often the ones whose work is published in lame alternative journals that merely claim peer review, and these are pretty easy for a seasoned researcher to spot. My conclusion is that when a study is of high quality and passes peer review to get into a reputable journal, it's generally reliable, regardless of who funded it.

So with that in mind, I plowed into PubMed to see what's being published about HFCS, careful to avoid studies funded by "Big Corn". I found quite a few clinical trials, like this one from the *American Journal of Clinical Nutrition* that concluded:

> Sucrose and HFCS do not have substantially different short-term endocrine/metabolic effects.

And this one, from the journal *Nutrition* that studied only normal-weight women:

> ...When fructose is consumed in the form of HFCS, the measured metabolic responses do not differ from sucrose in lean women.

Now I know from experience that I can't tackle a subject like this without being accused of being in the pocket of Big Corn, Big Pharma, Big Food, and Big Government as part of a giant evil conspiracy to keep everyone sick for fun and profit. I can stand on my head and shout "Don't overeat, eat healthier, avoid foods with added sweeteners" a hundred times, and people will still accuse me of saying everyone should go around with direct intravenous injections of HFCS. The culprit is overconsumption of high-calorie foods; singling out HFCS or replacing it with sugar does nothing to address health or weight problems. But, like I often say, you shouldn't listen to me anyway, you should find out the facts for yourself. Don't go to Big Corn or Big Government, and don't go to Oprah or other alternative healthcare promoters. Go to someone like the American Medical Association. They're not on anyone's payroll; they're an association of all of the world's best doctors, and their only purpose is to promote public health. Here's what the AMA has to say on HFCS:

> Because the composition of HFCS and sucrose are so similar, particularly on absorption by the body, it appears unlikely that HFCS contributes more to obesity or other conditions than sucrose... At the present time, there is insufficient evidence to restrict use of HFCS or other fructose-containing sweeteners in the food supply or to require the use of warning labels on products containing HFCS. The AMA ... recom-

mends that consumers limit the amount of added caloric sweeteners in their diet.

If you're overweight, stop overeating, and stop trying to place the blame elsewhere.

References & Further Reading

Hu, Frank. *Obesity epidemiology.* New York: Oxford University Press, US, 2008. 27.

Melanson K.J., Zukley L., Lowndes J., Nguyen V., Angelopoulos T.J., Rippe J.M. "Effects of high-fructose corn syrup and sucrose consumption on circulating glucose, insulin, leptin, and ghrelin and on appetite in normal-weight women." *Nutrition.* 3 Mar. 2007, Volume 23, Issue 2: 103-112.

Sizer, F., Whitney, E. *Nutrition: Concepts and Controversies.* Belmont, CA: Thomson Higher Education, 2007. 105.

Stanhope, K.L., Griffen, S.C., Bair B.R., Swarbrick M.M., Keim N.L., Havel P.J. "Twenty-four-hour endocrine and metabolic profiles following consumption of high-fructose corn syrup-, sucrose-, fructose-, and glucose-sweetened beverages with meals." *The American Journal of Clinical Nutrition.* 1 May 2008, Volume 87, Issue 5: 1194-1203.

USDA. "Foreign Agricultural Service." *Administering Sugar Imports.* United States Department of Agriculture, 15 Dec. 2009. Web. 10 Jan. 2010. <http://www.fas.usda.gov/itp/imports/ussugar.asp>

White, J. "HCFS: How Sweet It Is." *Food Product Design.* Virgo Publishing, LLC., 2 Dec. 2008. Web. 9 Jun. 2009. <http://www.foodproductdesign.com/articles/2008/12/hfcs-how-sweet-it-is.aspx>

36. The Mothman Cometh

The gravediggers were the first to see the strange being. It rose from the tree like a giant bird taking flight, but as it soared overhead, it began to look more like an angel, a man with wings. But it was dark and brown, more grotesque than radiant, more like a demon than an angel. The news spread.

He was seen again three nights later, when two young couples saw his red glowing eyes in the trees by the side of the road late at night. They drove faster but the strange dark man chased them, and they gave their story to the police. The news spread.

It was November of 1966 in the vicinity of Point Pleasant in the low country of West Virginia, and over the next ten days, the demon was seen at least four more times, either as a tall dark man prowling outside houses with his red eyes or flying overhead on leathery wings. By then the news had spread to the papers, and a reporter, probably Mary Hyre, writing for the newspaper *The Messenger*, inspired by the popular TV character Batman, consolidated all of these various reports into a single perpetrator that she dubbed the Mothman. But over time, as such things normally go, interest waned, the strange visitor stopped appearing, life resumed and the curiosity of the Mothman gradually fell out of memory.

In December of 1967, a little more than a year after the Mothman made headlines, the 40-year-old Silver Bridge connecting Point Pleasant with Gallipolis, OH across the Ohio River, collapsed under the weight of rush hour traffic. 46 people died. In such small towns, virtually everyone knew a few of the victims, and foolishness like the Mothman was quickly forgotten.

So why, more than 40 years later, do we still remember the Mothman? Why are there movies and books? Why is he considered a harbinger of disasters? He didn't hurt anybody or leave any evidence; there were only a few reports and they were unreliable at best. Yet the Mothman occupies a fairly high place among modern urban legends. We believe that when he appears, bridges collapse and people die, even though there was no such reference in any of the contemporary reports.

What do you do when you hear such a story? You really have very little to go on other than the popular legend. How much of it really happened? How do we know any of it really happened? Are you forced to accept it uncritically because you don't have these answers, or should you summarily dismiss the whole thing?

Well, as we see with so many of these stories that we examine, folk legends don't get very interesting until we introduce the element of an imaginative author with a book upon which fortune smiles. And so it happened with the Mothman. The author was John Keel, and the book was *The Mothman Prophecies*, published in 1975, nine years after the Mothman sightings. But although Keel was fortunate to have had his book turned into a movie, it wasn't the original. Much of the material was shared with Gray Barker's 1970 book *The Silver Bridge*, which was the first time a connection was drawn between the Mothman and the bridge collapse. Both authors came from the UFO genre, and both made suggestions the Mothman may have been an alien. Why an alien would choose the Mothman's behavior as the best way to make first contact with a new civilization is not convincingly argued in either book. Both authors peppered the skies of Point Pleasant with UFO appearances during the Mothman sightings, however I did not find any such accounts at all searching available local newspaper archives for that period.

Although popular accounts on the Internet state that the Mothman made as many as 100 appearances during the year leading up to the Silver Bridge collapse, this appears to be a modern fabrication. In the newspaper archives, I was only able to locate the half dozen or so reports in November 1966. That doesn't mean more didn't happen; it only means I couldn't find any evidence that anyone *at the time* reported that they did.

And those half dozen were not very well corroborated. The first episode, with the gravediggers, was in the town of Clendenin, 50 miles from Point Pleasant. Another was in Charleston, 45 miles from Point Pleasant. A couple of the reports are merely a dark-clothed man peering in through a window. Some of the reports have the Mothman flying on batwings, some of them have him merely standing around with reflective, animal-like eyes.

The Mothman's second sighting, by the four kids in the car, is the best known. They lived in the tiny town of Point Pleasant, WV, and drove late at night to the local "lovers' lane" called the TNT plant, a deserted explosives manufacturing and storage facility seven miles outside of Point Pleasant in the woods. While driving they passed a pair of red eyes by the roadside. They panicked and tried to get away, but the red eyes followed them, at speeds they reported of 100 mph. They reported to the local police, who said they knew the kids to be trustworthy witnesses, and that the eyes belonged to a man up to seven feet tall, with wings folded on his back.

The West Virginia Ordnance Works is kind of off by itself in the woods. It's not the kind of place you might happen to be driving by. There's nothing there except a grid of dirt roads in various states of being reclaimed by the forest. There are only a few scattered houses around, the nearest about a quarter of a mile away. Virtually nothing remains of the TNT plant and the area is now the McClintic State Wildlife Management Area, and has been since 1945 when the plant was decommissioned, decontaminated, and ceded over. There's actually more going on there now than there was in 1966, as it became a Superfund site in 1983 and they've been doing some cleanup since 2000. In 1966, it was basically just empty woods in the middle of nowhere. If people saw the Mothman there, they didn't just happen to glance outside. They had to make a deliberate trip to get out there. When a group of twenty-somethings who live in East Jesus drive out into the woods late at night, it's probably not stretching things too far to guess that alcohol may have been involved.

Investigator Joe Nickell has concluded that if the kids did see red eyes, they were probably those of one of the local owls called a Barred Owl. Combined with the lateness of the hour, the kids' state of mind, the speed of the car, and the recent news of the flying demon in Clendenin, Nickell is satisfied that the story has a plausible natural explanation. And, of course, the story is anecdotal only, and has its own credibility issues for whether anything took place at all, owl or not. Certainly 100 mph is highly suspect, judging by a glance at the roads on Google Earth.

Was this alleged nighttime chase truly an unambiguous prophecy of destruction to come?

Trying to connect the Mothman's appearance with the Silver Bridge collapse seems like a bit of a shoehorn job to me. They happened many miles and a full year apart. If you were a Mothman and wanted to foretell a bridge collapse, wouldn't you choose to appear closer to the bridge, or at least sooner before it happened? It's not even great fiction for Barker and Keel to have conceived. I'm sure lots of things happened in Point Pleasant during that intervening year, a lot closer to the bridge, and any of those events would make a more compelling candidate for an omen.

The Mothman made the news more recently when the I-35W bridge over the Mississippi in Minneapolis, MN collapsed in 2007, killing 13 people. It wasn't until *after* the bridge collapse that people on Internet forums, and in particular a caller to the Coast to Coast AM radio show, reported that the Mothman had been seen in Minneapolis prior to the collapse. Such reports are highly suspect, since there are no published reports of this at all, and people were simply regurgitating the

Silver Bridge incident and *The Mothman Prophecies* movie. Any connections you may have heard between the Mothman and this later bridge collapse are pure *ex post facto* fantasies.

Barker and Keel had both come to Point Pleasant to write UFO books. While they were watching the skies for UFOs and writing imaginative tales about Men In Black and government conspiracies and mysterious phone calls and strange warnings, stories told by thrillseeking kids and ladies who hadn't closed their curtains all the way stole the headlines. And later, a real disaster — that cost real lives and plummeted everyone in town into mourning — brought the newspapers back down to Earth. Barker and Keel had little choice but to include these events in their books. A frequent criticism of the movie *The Mothman Prophecies*, in which the filmmakers tried to condense Keel's book into something like a tellable story, is that there is no apparent connection between practically any of the events in the film. And that's perhaps the best way to conclude a skeptical examination into the story of the Mothman: A few vague reports, probably not connected; none of which had anything to do with the Silver Bridge's structural collapse more than a year later.

References & Further Reading

Barker, Gray. *The Silver Bridge*. Clarksburg: Saucerian Books, 1970.

Hyre, Mary. "Winged, Red-Eyed 'Thing' Chases Point Couples Across Countryside." *Athens Messenger*. 16 Nov. 1966, Vol 61, Number 271: 1.

Keel, John A. *The Mothman Prophecies*. New York: Saturday Review Press, 1975.

Nickell, Joe. "Mothman Solved." *Skeptical Inquirer Mailing List*. Committee For Skeptical Inquiry, 31 Jan. 2002. Web. 23 Mar. 2008. <http://csicop.org/list/listarchive/msg00317.html>

Phillips, J., Jensen, H. "Loaded with Cars, Big Span to Ohio Collapses in River." *Charleston Daily Mail*. 16 Dec. 1967, Volume 149, Number 145: 1-3, 6.

37. Sarah Palin is Not Stupid

Today we're going to delve into the minds of those who actively promote misinformation, political oppression, terror, conspiracies, and anything else that detracts from the public good. What drives them to do so? Are they right in their own minds, or do they know that what they do is wrong? More importantly, what should we know and understand about these people? I'm going to go out on a limb and start with a concept that may seem shockingly politically incorrect to some: I'm going to disagree with the popular perception that Sarah Palin is nuts.

Let me tell you something about Sarah Palin, but first with the understanding that I don't know any more about her than you do; I've never met her either; and I didn't vote for her. Stupid people don't tend to attract contributors, managers, supporters, and electorates. If she'd exhibited stupidity on the Wasilla city council, they probably wouldn't have elected her mayor. If she'd exhibited stupidity as mayor, they probably wouldn't have elected her for a second term. Her appointment to the Oil and Gas Committee by the governor was probably not because she'd behaved stupidly. Finally, stupidity probably does not characterize most successful bids to run for governor of one of the United States. Does she exhibit an almost robotic and uncritical point-by-point support of the Republican platform? Yes. Is she undereducated for her position? Possibly, she has a bachelor's degree in journalism. It's arguable that she's demonstrated a clear disdain for, and illiteracy in, science. She gives every indication that her religious beliefs and her party guidance determine her priorities. But welcome to reality: That's the way a lot of people work, including a lot of people on the other side of the political aisle.

If you call yourself a critical thinker, ad hominem attacks should not be the extent of your criticisms of those in whom you find fault. Show me one thing Sarah Palin has said or done that's "stupid", and I'll show you something that's perfectly rational for someone with her religious and political convictions. It may be that you simply disagree with her convictions, and you probably have very good reasons for doing so. But if that's the case, don't just say "Sarah Palin is stupid". That's kindergarten talk, and it makes you look bad, not her. Understand why she takes the position she does, then reveal the faults in that position.

My point today has nothing to do with Sarah Palin, or with anyone else. It has to do with a lack of critical thinking among many people who consider themselves skeptics. A lot of prominent people are dismissive of science: Celebrities, politicians. Many of us tend to dismiss them right back as irrational or nuts. But this demonstrates exactly the same kind of shortcutted thinking that we're accusing them of.

For example, I heard some skeptics the other day talking about Bill Maher, saying "I didn't realize he was as crazy as he is." (Bill Maher is an outspoken critic of science based medicine. He's endorsed AIDS denialism, Big Pharma conspiracies, anti-vaccination, and natural medicine.) Now, granted Bill Maher is wrong about a lot of things, but he's not on the fringe. A lot of people believe that stuff. Clearly it's important that they be educated, because widespread beliefs like this would represent a serious national health crisis. If you dismiss those beliefs as craziness, you're saying there's nothing to them, they're meaningless. Instead, acknowledge that there are compelling cultural influences that have led Bill Maher and others to believe those things. Bill Maher is just one of many victims of these influences, and it's because he has the average person's ability to understand and interpret the information he's been exposed to, not because he's crazy.

In the same way, you could say Sarah Palin is simply responding to cultural and political influences. People need cheap energy, so she's a proponent of drilling the oil in her state. People want government to eliminate wasteful spending, so she bashes fruit fly research, the significance of which has never been made clear to her or to the public. The United States is a strongly Christian nation, and many people support teaching creationism in schools, and oppose stem cell research. Palin isn't being stupid by embracing these concepts, she's responding to the same influences everyone else is.

If you were to sit down and have a conversation with Mahmoud Ahmadinejad, in such a way that you could both speak your native language in order to be articulate and insightful, I bet you'd find that he's knowledgeable, well spoken, and intelligent. (Maybe I'm wrong, I haven't had such a conversation with him, but from what I've read of his background I'd say it's practically a certainty that he has his act together pretty well.) Unless you happen to be a Muslim fundamentalist, you're likely to disagree pretty strongly with many of his beliefs and priorities. But I bet he'd convince you that his convictions run pretty deep, and have solid historical and cultural roots that are not going to be washed away from an entire nation overnight. I bet you'd say "Wow, he actually has a point of view, and what he's saying makes sense within that context." This characterization is acutely different from the whimsical ramblings of a nut. If you dismiss Ahmadinejad as a crazy whackjob, not only are you factually wrong, but you do it at your peril, because you are grossly underestimating the depth and foundation of what you're protesting against.

I've watched two Muslim executions by stoning, in all their graphic detail — on video, I hope I never see it in person — because we live in a world where this actually happens, and I feel it's important to understand what I object to as fully as possible. It's easy to watch a stoning and conclude that only a crazy person could willingly be a part of such a medieval horror. What's just as frightening as the stoning itself is that the people doing it are someone's nextdoor neighbors. They take their kids to the park. They give birthday presents. They paint and write and play musical instruments. They are, in fact, quite human. And yet they're capable of something that's unthinkable to you or I. It's not because they're crazy. It's because they're smart people who are profoundly dedicated to their belief system, and who were raised in a frame of reference that lets them stone a person to death with the same regard as a Westerner might kill an enemy in battle. It's a necessity, it's a duty, and it's the right thing to do. If you dismiss these people as crazy or as zealots, you are factually wrong, you're missing the point, and you're failing to understand what it is you object to.

Look at Timothy McVeigh, the Oklahoma City bomber, and consequently the United States' greatest mass murderer of children. To best prepare ourselves to prevent this kind of thing happening again, we have to be sure that we accurately understand the motivations behind it. McVeigh is a guy who lived in a world of conspiracies. The people he surrounded himself with all believed the same thing: That the government was out to get them. When you live and breathe that 24 hours a day, when it's your entire sphere of influence, it's not delusional. It was a vicious circle. The more input he received, the more he sought out such information. Well-understood perceptual phenomena like confirmation bias made it normal and healthy for McVeigh's brain to reject information that did not indicate the government was out to get him. Eventually he got to a point where the best move — in the context of what he believed was going on — was to strike back, as violently as possible. We are better prepared to deal with Timothy McVeighs if we don't allow ourselves the intellectually lazy shortcut of "Oh, he was just some nut."

The same goes for Sarah Palin, Ben Stein, Ken Ham, Bill Maher, Jenny McCarthy and Jim Carrey, and Prince Charles, all people who actively promote bad science or misinformation, and who believe they're doing the right thing. That's an important point that's too often overlooked. With few exceptions, most honest promoters of bad information have good intentions. They're not crazy raving lunatics out to get us. If you want to have an informed, rational conversation with one of these folks, and you want them to be receptive to your statements, approach them as you would any public figure who works hard in the public good. At a fundamental level, they're on our same team: They want what's best for people.

Prince Charles is a nutcase who has no idea what he's talking about.

...makes *you* sound like a close-minded radical, making irrational ad hominem attacks.

Prince Charles is a good man who cares deeply for the public welfare. Unfortunately, a lot of the medical information he passes along is woefully out of date.

...has a chance that someone will actually listen to it. And expressing yourself in a way that's worthy of people listening is an important, but all too often overlooked, part of the promotion of critical thinking.

REFERENCES & FURTHER READING

Alexander, Y., Hoenig, M. M. *The new Iranian leadership: Ahmadinejad, terrorism, nuclear ambition, and the Middle East.* Santa Barbara: Greenwood Publishing Group, 2008. 30-35.

Booker, C., North, R. *Scared to Death.* London: Continuum UK, 2007.

McKay, Charles. *Extraordinary Popular Delusions and the Madness of Crowds.* NY: Universal Digital Library, 1841.

Michel, L., Herbeck, D. *American Terrorist: Timothy McVeigh & the Oklahoma City Bombing.* Darby: Diane Pub Co., 2003.

Shermer, Michael. *Why People Believe Weird Things.* NY: Henry Holt and Co., 2002.

Sternberg, Robert. *Why Smart People Can be So Stupid.* New Haven: Yale University Press, 2002.

Thompson, Damian. *Counterknowledge.* NY: WW Norton and Co., 2008.

Van Hecke, Madeleine. *Blind Spots: Why Smart People Do Dumb Things*. NY: Prometheus Books, 2007.

38. Locally Grown Produce

Today we're going to be politically incorrect again and point our skeptical eye at another sacred cow: Locally grown produce. Particularly in the United States, but in many other countries as well, one of the newest and fastest growing market segments is locally grown produce. The claims are that locally grown produce is less wasteful of fuel because it doesn't need to be delivered over long distances; it's fresher for the same reason; and it supports a small local organic farmer instead of an immoral megacorporation that sources food from cheap overseas producers.

I discussed one of these claims, about local delivery burning less fuel, in a May 2009 entry on SkepticBlog.org. It must have been pretty inflammatory, because it generated a huge number of comments. Most of them followed this pattern: The commenter begrudgingly agreed with the mathematics of the delivery question, but then claimed that I missed the point completely because the real reason to like locally grown produce has nothing to do with a low carbon footprint of minimal delivery miles. I'm not sure I buy that — virtually everyone I've ever asked says that's what locally grown is all about — but hey, I'm fair, we'll give them all a voice here.

First, let's give a brief overview of the mathematics of local delivery. Think of the traveling salesman problem. This is where you speckle a map with all sorts of random locations. The traveling salesman's problem is to find the shortest possible driving route, called a tour, that visits each of the locations. It's among the most computationally difficult problems in mathematics. But there's a cool piece of free software by Michael LaLena that finds one efficient solution using a genetic algorithm. Try to stump it with a pattern of hundreds of dots that you think will be hard to connect, and the software blows your mind with a surprisingly simple tour that visits all the locations.

Many years ago I did some consulting for a company that was then called Henry's Marketplace, a produce retailer built on the founding principles of locally grown food. Henry's had evolved from a single family fruit stand into a chain of stores throughout southern California and Arizona that sold produce from small, local farmers. Part of what I helped them with was the management of product at distribution centers. This sparked a question: I had assumed that their "locally grown produce" model meant that they used no distribution centers. What followed was a fascinating lesson where I learned part of the economics of locally grown produce.

In their early days, they did indeed follow a true farmers' market model. Farmers would either deliver their product directly to the store, or they would send a truck out to each farmer. As they added store locations, they continued practicing direct delivery between farmer and store. Adding a store in a new town meant finding a new local farmer for each type of produce in that town. Usually this was impossible: Customers don't live in farming areas. Farms are usually located between towns. So Henry's ended up sending a number of trucks from different stores to the same farm. Soon, Henry's found that the model of minimal driving distance between each farm and each store resulted in a rat's nest of redundant driving routes crisscrossing everywhere. What was intended to be efficient, local, and friendly, turned out to be not just inefficient, but grossly inefficient. Henry's was burning huge amounts of diesel that they didn't need to burn. So, they began combining routes. This meant fewer, larger trucks, and less diesel burned. They experimented with a distribution center to serve some of their closely clustered stores. The distribution center added a certain amount of time and labor to the process, but it still accomplished same-day morning delivery from farm to store, and cut down on mileage tremendously. Henry's added larger distribution centers, and realized even better efficiency. Today their model of distributing locally grown produce, on the same day it comes from the farm, is hardly distinguishable from the model of any large retailer.

Compare the traveling salesman's simplified tour to a tangle of crisscrossing bicycle spokes, and the inefficiency of direct delivery between farm and store becomes acutely clear. If we want to minimize the carbon footprint of the entire food cycle, eliminating direct delivery is the easiest place to make the biggest gains. So, right off the bat, the main reason most people prefer locally grown produce is shot down, and shot down in big flames. But let's turn to the SkepticBlog commenters and see what people had to say.

As did a number of readers, Ian pointed out that you have to consider the total price. Not just the cost of distribution, but also the cost of the retailer's wholesale purchase. Total them all up, and in some cases it might be cheaper to buy from ridiculously far away:

> ...Wal-Mart [buys] fruit from South Africa, coffee from Kenya, etc. Flying this produce around the world is clearly using more fuel than even an inefficient model for distributing food locally. The efficiency comes not from reducing fuel usage, but from paying significantly less for the produce.

This was underscored by another poster, "Old White Guy":

> As someone who spent a good chunk of his life controlling distribution for several large companies, I can say the only thing that matters is getting the product to the point of sale as inexpensively as possible. If that [means] the cheapest wine in the store comes from another continent, so be it.

This suggests that it some cases, huge container-sized purchases might still be cheaper for the large retailer, even though their delivery produces a lot of wasteful emissions, and their production might be with some god-awful third-world high-pollution child-labor dogs-and-cats-living-together environmental disaster. That might be true in some cases, but those would be the exception, not the rule. Most of the time, produce is cheaper from those countries because the native growing conditions are much better for that particular crop. Tomatoes flourish in Spain but require heated greenhouses in the United Kingdom, and so the overall energy efficiency of growing them in Spain and transporting them overseas to the UK is actually better.

A number of people who disagreed with my article repeatedly referenced Michael Pollan's book *The Omnivore's Dilemma*. Pollan devotes one of the book's four sections to the practices of holistic cattle farmer Joel Salatin. One of Salatin's rules is that, in the interest of a minimum carbon footprint, he won't ship his beef at all; customers have to drive to him to pick it up. While I applaud Salatin for having the right idea and the right motivations, I don't believe he thought through this particular point very critically. Salatin should instead design practices that more directly address his desire: He should allow only shipments that use a minimum amount of fuel per pound of beef delivered. Instead, he adopts a rule that might put hundreds of cars and vans on the road, each delivering only a few pounds of beef. Salatin's solution is emotionally satisfying and makes for a fine sound bite, but its underlying science is flawed and counterproductive to his stated goals.

The elephant in the room on Joel Salatin's farm is that his near-total self-sufficiency methods require an outrageous 550 acres to support only 100 head of cattle and a herd of pigs, plus some turkeys and chickens. Most of the acres are used to grow the feed and raw materials the animals require. I didn't find any valid defense of this, and Pollan's book simply avoids the issue. Typically, pasture-fed cows require half an acre each (a number which varies with climate), so Salatin is using about *ten times* as much land as he should. Such wasteful land usage might work well in the case of a high-end boutique retailer like Joel Salatin, but it's clearly well beyond the limits of practicality for the world's real food needs.

The overall picture is often a lot more complicated than simply "locally grown". Let's say you want sheep or dairy products, and you live in New York. Where are those products going to come from? Certainly not from anywhere local. If you get them from a state or two away, which is about as local as possible, what went into their production? A lot of feed, for one thing. But spin the globe and look at New Zealand. New Zealand has the world's most efficient sheep and dairy industries, and one big reason is their climate and conditions that allow year-round grazing. According to the New York Times:

> *Lamb raised on New Zealand's clover-choked pastures and shipped 11,000 miles by boat to Britain produced 1,520 pounds of carbon dioxide emissions per ton while British lamb produced 6,280 pounds of carbon dioxide per ton, in part because poorer British pastures force farmers to use feed. In other words, it is four times more energy-efficient for Londoners to buy lamb imported from the other side of the world than to buy it from a producer in their backyard.*

And yet many of the same people who are so vocal about a minimum carbon footprint consider this massive net energy savings to be immoral because it includes overseas transport. Why? Is it a geopolitical preference? Is it a matter of supporting farms from your own country instead of sending money overseas? OK, fine, that's an absolutely valid point of view. But if your true motivations are political, don't greenwash them and claim that you're really interested in environmental science.

If it's support for small business, if you'd rather support someone like Joel Salatin than a megacorporation like Wal-Mart, that's also an absolutely valid point of view. Just call it what it is instead of greenwashing it and claiming environmental awareness. To get the premium boutique experience, Salatin's customers burn way more gas per pound of beef delivered than do Wal-Mart's container ships from New Zealand. If you have other reasons to object to Wal-Mart's New Zealand beef, fantastic; just be aware of what your objections really are. It's more intellectually honest, it's more insightful, you'll learn more, and you're not being disingenuous.

Don't get me wrong, I love farmers' markets. We go to our local one sometimes and it's a fun family event for us. We love the giant, wonderful tomatoes and strawberries that you can't get at the supermarket. But I understand that farmers' markets are more of a community experience than an efficient (or "green") way to buy food. The real reasons to enjoy your farmers' market have nothing to do with it being somehow magically environmentally friendly. Too often, environmentalists are satisfied with the mere appearance and accouterments of environmentalism, without regard for the underlying facts. Apply some mathematics and some economics, and you'll find that, more often than not, a smaller environmental footprint is the natural result of improved efficiency.

References & Further Reading

Gutin, G., Punnen, A. *The Traveling Salesman Problem and Its Variations.* Dordrecht: Springer, 2002.

McWilliams, James E. *Just Food: Where Locavores Get It Wrong and How We Can Truly Eat Responsibly.* New York: Little, Brown and Company, 2009.

Saunders, C. Barber, A. Taylor, G. "Food Miles - Comparative Energy Emissions Performance of New Zealand's Agriculture Industry." *AERU Research Report series.* 1 Jul. 2006, Research Report, No. 285.

USDA. "Cost of Food Services and Distribution." *USDA.* United States Department of Agriculture, 29 Aug. 2000. Web. 29 Aug. 2005. <http://usda.gov/news/pubs/fbook99/sections/1b.pdf>

Weber, C., Matthews, H.C. "Food Miles and the Relative Climate Impacts of Food Choices in the United States." *Environmental Science and Technology.* 16 Apr. 2008, Volume 42, Number 10: 3508-3513.

39. How to Make Skepticism Commercial

Today we're going to answer the question that's had a lot of skeptics scratching their heads for a while, myself included. I'm going to show how to make the promotion of skepticism both profitable and popular. You may not like the answer, because it requires a lot of hard work; but it's simple and it's doable.

There are many of us who spend quite a lot of time trying to think of ways to make the promotion of critical thinking profitable. The people who sell bad and even harmful information make lots of money, because they're selling easy answers. But, try to sell good information that will actually improve peoples' lives, and nobody will buy it. Skepticism is the opposite of what people want to buy. It's the opposite of what characterizes an attractive product. We're not selling magically easy answers to real problems, and so nobody wants to listen to what we have to say.

And so, we're left toiling away, making podcasts, writing books like this one, giving lectures, having to give most of it away for free, and reaching only a tiny audience who is already predisposed to hear our message. It's clear that what we've been selling so far has not been successful in the marketplace.

But of course, you might argue that commercial success is not what we're going for. We're not here to make money, we're here to teach people to think better so they can make better decisions and lead happier, healthier lives. It's a public service, not a money making scheme. I say the two go hand in hand. If a message can be made commercially desirable, it's going to reach a lot more people. People will want to hear it. We need to deliver the message of critical thinking in such a way that it enters the pop mainstream. A desirable message that people clamor to pay for has a thousand times the reach of a message that must be given away for free.

And here we've sat, rubbing our chins and trying to solve this conundrum. Many times I've had lunches or Skypes with other skeptics and we've wondered how to bend the edges of the skeptical message into something popular. How do you take the opposite of what's popular and make it popular, without changing it? And nobody's ever had a clue. But then, one day not long ago, the clouds opened, I heard an angelic chorus, and the answer came to me.

Let's begin by establishing a premise. Let's understand the characteristics that make a popular product desirable. If we're interested in television programs, we should look at the popular shows that promote bad information and understand what makes them popular. I argue that it's characters, storyline, and excitement that make them popular, not the bad information. Audiences demand a show that entertains them; they don't demand that shows contain bad information. A skeptical TV show should no more be pitched as presenting good information, than a show about psychics should be pitched as presenting bad information. *The quality of the information given in a show is not its selling point.*

If audiences aren't demanding shows that present bad information, like psychics and ghosts and magical cures, why do networks tend to respond with so much programming filled with it? Simply because psychics and ghosts and magical cures are an easy, low-budget way to get the Wow! factor and make a show compelling. Nobody at the network is twisting their mustache and plotting ways to spread harmful information to an unsuspecting public, and no TV viewers are writing in and asking for shows to further reduce their scientific IQs.

Hollywood has a term for the mechanical plot element around which a story is built. Describing the meaning of the term 'McGuffin', Alfred Hitchcock said "In crook stories it is almost always the necklace and in spy stories it is most always the papers." But the actual story you're enjoying is about Cary Grant and Grace Kelly. The McGuffin is there, but it's not the main point and it's not what makes the movie good. To insert a message of critical thinking into a movie, it needs to be like the McGuffin. You must deliver a good, exciting, engaging story first; and within that story you can weave the importance of critical thinking. But if you set out to make a movie *about* critical thinking, you set yourself up for failure.

If you're talking about news outlets that badly distort stories in order to add zing and make headlines like *Scientists Baffled by Image of Jesus in Tree Stump*, the cause is exactly the same. The newspaper's motivation is not to promote bad information, it's to entertain and sell papers. Promotion of bad information is simply one easy way to do that. The editors aren't deliberately trying to ensure that the information is wrong any more than the readers are demanding that they get more stories wrong.

Look at bad information in business, like a multilevel marketing scheme built around some health product. The products are usually overpriced and worthless, and the business models can only succeed for the people at the top selling the product, never the distributors required to buy the product. To get good salespeople, you have to pay for them: That's expensive and good people are hard to find. It's an easy, seductive alternative to simply deceive people into thinking they'll become millionaires if they buy your product at the special "distributor price" and pass that savings along to their friends.

So we have people promoting bad information in entertainment, news, business, and other walks of life, simply because the supernatural, the miraculous, and the too-good-to-be-true is an easy way to generate interest. When your job is to generate interest, you stick with what works. The inevitable result is that the population is continuously reinforced with a pattern of thinking founded upon faulty beliefs, and that leads to decision making based on wrong information. For any product of any type to deliberately promote bad information as a lazy shortcut to make it interesting, is exactly like promoting high-nicotine cigarettes as an easy way to get a cheap high. It's lazy, and it's ultimately harmful.

Every one of those TV documentaries, news programs, supermarket supplements, and business models could be made just as interesting, as attractive, and profitable if they put in the extra work and founded their product upon sound information. People often ask if we should blame the producers who make the bad documentaries or the audience who demands them. I have to lay the blame entirely upon the content decision makers. The audience is demanding only to be entertained; it's the decision makers who are choosing to do that through the promotion of misinformation. I also blame the supplement makers. Customers ask only to be healthy; it's the supplement makers who choose to deceive customers with a worthless product that claims to confer super health. Neither the documentary audience nor the supplement customers want their product to be based on information known to be wrong.

So when we, the promoters of critical thinking and sound evidence, ask the question "How do we make it profitable to sell good information?", there is no magic formula. There is no paradigm shift waiting to be discovered in some late-night beer-fueled epiphany. I will not sell you an easy answer. The solution is to work harder than the people who shortcut.

- If you are making a TV documentary, go the extra mile and blend correct information with dramatic storytelling. Do not take the easy way out and suggest that ghosts might be the answer.

- If you are writing a newspaper article, go the extra mile and find what's really exciting about the actual facts. Don't cheat with a misleading headline or go for some lame "human interest" angle that focuses on someone's incorrect beliefs.

- If you are looking for a product to sell, work hard and invent a truly good product. Don't shortcut and sell a worthless health supplement or some imported plastic crap that you can lie about in an infomercial. Go the extra mile to design a distribution model based on sound principles and hard work, rather than taking the easy way out and planning to deceive people into believing they'll become millionaires joining your multilevel marketing network.

Oprah Winfrey and Kevin Trudeau have become multimillionaires because they grab every piece of low hanging fruit. Oprah finds it easy to get ratings by promoting everything that's interesting without putting it through the reality filter. Kevin Trudeau finds it easy to sell books by playing on peoples' most obvious desires, to achieve super health and super wealth effortlessly. I argue that it is possible to be as successful as Oprah and Trudeau by promoting good information that actually helps people, but you would have to work ten times, or 100 times, as hard as they do. But if you do put in that work, you could indeed have the #1 show on television and the New York Times #1 bestseller that promote sound critical thinking. You just have to find a way to wrap it in good entertainment that people want.

Skepticism is just the McGuffin. It is possible to successfully sell skepticism, but you must do it by delivering it within something appealing that people want. I'll close with my favorite quote from *The Amazing Meeting 7*, from Jennifer Ouellette of the Science and Entertainment Exchange: If you want to reach for peoples' minds, reach for their hearts.

REFERENCES & FURTHER READING

Bardi, Jennifer. "The Humanist interview with Neil Degrasse Tyson." *The Humanist.* 1 Sep. 2009, Volume 69, Number 5: 9-11.

Herreid, Clyde Freeman. *Start with a story: the case study method of teaching college science.* Arlington, Vancouver, Canada: NSTA Press, 2006.

Hubbard, Rob. "How do we create engaging and effective learning content?" *MindMeister*. MeisterLabs, 25 Nov. 2009. Web. 25 Jan. 2010. <http://www.mindmeister.com/en/maps/show_public/18195804>

Olson, Randy. *Don't Be Such a Scientist*. Washington, DC: Island Press, 2009.

Ouellette, J. *The Physics of the Buffyverse*. New York: Penguin Group, 2006.

Stafford, B. *Artful Science: Enlightenment Entertainment and the Eclipse of Visual Education*. Cambridge: The MIT Press, 1996.

40. What's Up with the Rosicrucians?

What do you get when you mix alchemy, *The Da Vinci Code,* Nazis, Christianity, mysticism, the Knights Templar, Shakespeare, *The Secret,* and ancient Egypt? No, not a bad movie about Ben Stiller working late at a museum; you get the Rosicrucians. Who are they, what are they up to, what do they believe, and what the heck's the deal with all the historical imagery?

In San Jose, California, stands an Egyptian obelisk, covered in hieroglyphics. Nearby is a statue of Caesar Augustus, outside a planetarium in classical Islamic architecture. In the midst of this historical montage, surrounded by living papyrus plants, is the Rosicrucian Egyptian Museum, actually quite a good museum filled with authentic Egyptian artifacts. The rest of this city block is taken up by the world headquarters of AMORC, the Ancient Mystical Order Rosae Crucis. The name Rosicrucian comes from Rosy Cross, an ancient symbol that's been adopted by many religious and pagan groups throughout history. To the modern Rosicrucian organization, the cross with an unfolding white rose in the center represents the human body and its consciousness opening up, carefully steering away from its more common traditional connections with Christianity. The Rosicrucians downplay any religious associations with their symbology, claiming not to be a church, and welcoming members of any religion or no religion. (Here's a hint: When you're taking peoples' money, don't turn anyone away at the door.)

According to tradition, the founder of Rosicrucianism was the none-too-improbably named Christian Rosenkreuz, born in 1378, the last surviving member of an assassinated German noble family, secreted away to a monastery where he grew to found the order that bore his name. Rosenkreuz traveled throughout the Christian, Muslim, Dharmic, and pagan lands, amassing his knowledge and acquiring a small but tight group of followers. Of his death, all that is known to Rosicrucian tradition is that his body lies somewhere in a geometrically proportioned cave, incorrupt, and bathed in white light from an unseen source.

Rosenkreuz's story is told in the *Fama Fraternitatis Rosae Crucis*, an anonymous manifesto published in Germany in 1614. The following year, another manifesto appeared, the *Confessio Fraternitatis*, which declared the existence of a secret society of alchemists and sages following pious Christian principles and planning an intellectual enlightenment of Europe. Then in 1616, the third and last of the Rosicrucians' three major manifestos was published, *The Chymical Wedding of Christian Rosenkreutz*, an allegorical tale of Rosenkreuz using alchemy to assist in the wedding of a king and queen in a strange and magical castle. The three manifestos made quite a splash in certain circles. Leaders of the occult and science tried to make contact with the secret society described, including Rene Descartes, William Shakespeare, and the philosopher and scientist Sir Francis Bacon. In fact, by some accounts, Francis Bacon was not only actually one of the secret society members, he may have written the first two manifestos; and some Rosicrucians claim he wrote Shakespeare's works as well. Another hint is that Bacon was also a member of a Templar society, and the Knights Templar bore the same rose-colored cross as the Crusaders. Some believe the third manifesto was written by the Lutheran alchemist Johannes Valentinus Andreae, whose name was also claimed in a 1960's hoax as one of the Grand Masters of the Priory of Sion, which figured so prominently in *The Da Vinci Code*.

So suffice it to say that there is enough pop-culture quasi-history to adorn Rosicrucianism with as much illustrious intrigue as you wish. Our task is to see if we can connect the dots, and find out what links there are, if any, between all those legendary characters and the people who sit in offices in San Jose today, depositing checks and doing the books. Exactly what are they up to? What do they do, and what do Rosicrucian members do? Here's the answer.

If I were to summarize the modern Rosicrucian organization, I'd compare it to a low-pressure, less expensive version of Scientology, based on New Age beliefs instead of L. Ron Hubbard's science fiction. You send them a few hundred dollars a year for your membership, and they send you printed lessons for self study that teach you all about their mystical belief system, the "keys to universal wisdom", as they put it. Like Scientology and Freemasonry, Rosicrucians reach various levels, or degrees, based on how much of the self-study material you've purchased and read. You can even perform your own initiation ceremonies into each new degree at home. In your first five years as a Rosicrucian, you'll cover the three "neophyte" degrees from First Atrium through Third Atrium, and then the "temple" section from First Temple Degree through Ninth Temple Degree. By this time your teaching will include topics such as:

- Mental Alchemy
- Telepathy, Telekinesis, Vibroturgy, and Radiesthesia
- Cosmic Protection, Mystical Regeneration
- Attunement with the Cosmic Consciousness

One of the benefits available to modern Rosicrucians is magical assistance to those in need of actual assistance, which they provide to successful petitioners via their "Council of Solace". Their web site describes how this works:

> *The Council does this by putting certain spiritual energies into motion and directing them in accordance with mystical law and natural principles. Metaphysical aid is thus directed to individuals ...with health, domestic, economic, or other problems, and aid is also directed to those who are attuned with the Council. The aid of the Council of Solace operates on the cosmic plane. Its activity is solely metaphysical and in no*

way interferes with any professional or health-care assistance being received on the physical plane.

So at this point you're probably yawning at this yet-another "spin the wheel and invent a New Age philosophy". So it's a good time to introduce William Walker Atkinson, an author who wrote about 100 books in the early 20th century under many pseudonyms. He is credited with being one of the principal architects of the New Thought movement, which evolved into today's New Age movement. His book *The Law of Attraction in the Thought World* is one of the primary influences of Rhonda Byrne's book and movie *The Secret*, and in fact the word "Rosicrucian" appears subtly on screen throughout the movie's title transitions. Many of the principal writings of the Dharmic movement of the 1960's, so popular with the Beatles and attributed to various swamis and yogis, were in fact written by Atkinson. But one of Atkinson's books broke the pattern and was written not to promote the New Thought mysticism, but rather to expose it. Published under the name Magus Incognito, its title was *The Secret Doctrine of the Rosicrucians*. In it, Atkinson claims that the *true* Rosicrucian order does not accept fees, has no formal organization, and is in fact secret. He then gives away all the contents of the Rosicrucian degrees. Why would he write this book?

AMORC, the modern formal Rosicrucian group, was launched in New York in 1915. The original founder, Harvey Spencer Lewis, and its first leader (or "Imperator" as they call it), is said to have borrowed quite heavily from the works of Yogi Ramacharaka in developing the Atrium and Temple Degree series. Who was the real author behind the name Yogi Ramacharaka? You guessed it, William Walker Atkinson. Apparently annoyed that his work had been so broadly and obviously "borrowed from" (to put it politely) without attribution, Atkinson quickly produced *The Secret Doctrine of the Rosicrucians* by retitling some of his own earlier works that contained the material used in the Rosicrucian lessons, and adding a few jabs like "real Rosicrucians would never take your money the way AMORC does".

Atkinson also reminded us that the term Rosicrucian and the rosy cross symbol have both been in the public domain for centuries, so nobody has any exclusive right to use them; and in fact that there are many competing Rosicrucian groups out there. Although AMORC has clearly won in the marketplace with its expansive San Jose headquarters, you might also choose to join the Ancient Order of the Rosicrucians, the Fraternitas Rosicruciana Antiqua, the Lectorium Rosicrucianum, or any of a dozen others, all based on essentially the same occult New Age mystical traditions.

Ever since the original manifestos were published by the first in this long line of clever authors, it seems everyone's been trying to get in on the Rosicrucian action; either directly by name or by rebranding it the way Rhonda Byrne, and in fact William Atkinson himself, have done. It's even been borrowed by whole nations in search of a defining philosophy. In his book *The Occult Roots of Nazism,* author Nicholas Goodrick-Clarke found that Nazi symbology was inspired by an 18th century German Rosicrucian order called *Gold und Rosenkreuzer.*

And thus we have a ten-cent tour of the history of Rosicrucian mysticism. It was invented in the early 1600's by European intellectuals who wrote allegorical tales blending alchemy with Protestant Christianity. It was revived in the early 1900's by the New Thought movement seeking ancient forms of mysticism that appealed to the notions of a population just beginning to learn that such a thing as a cosmic universe existed, and searching for meaning within it. And a century later, Rosicrucianism remains just one more flavor of for-profit New Age products, leveraging claims to ancient wisdom into bank deposits. It professes that the "keys to the universe" were known to a handful of Europeans 400 years ago, they just never managed to do much with them, since recurring credit card billing hadn't been invented yet.

I will close with the phrase that Rosicrucians like to put at the bottom of all their written communications. It means "So it shall be" and is often used to mean "Amen" or "In the name of God":

So Mote It Be!

References & Further Reading

AMORC. "Mastery of Life." *Rosicrucian Order*. Ancient Mystical Order Rosae Crusis, 1 Jan. 2009. Web. 13 Jan. 2009. <http://www.rosicrucian.org/about/mastery/>

Atkinson, W. *The Secret Doctrine of the Rosicrucians*. Chicago: L.N. Fowler & Co., 1918.

Editors. "Rosicrucians." *Columbia Electronic Encyclopedia*. Columbia University Press, 1 Oct. 2009. Web. 12 Jan. 2009. <http://www.infoplease.com/ce6/society/A0842439.html>

Schwarz, Avraham. *Empowering Thoughts: The Secret of Rhonda Byrne or The Law of Attraction in the Torah*. New York: BN Publishing, 2007.

Yates, F.A. *The Rosicrucian Enlightenment*. London: Routledge, 1972.

41. Real or Fictional: Food and Fashion

Today we're going to open the pages of pop culture to the chapter on American products named after people. People whose names you've grown to love and trust. But which of those are names of real people who actually existed, and which are the inventions of marketing professionals? You will probably know some of these, but you won't know all of them. See how many you can get right.

Let's begin in the packaged food aisles of the supermarket. There are a lot of names in here. Some of them are just characters invented by the product marketers to present a wholesome, homey image; but how can you tell those apart from the real names of food company founders from a century or more ago? Here's an easy one to get us started:

Pancake matron **Mrs. Butterworth:** Fictional. You probably guessed that her name was just a little too improbable. Mrs. Butterworth is represented in commercials by a talking syrup bottle in the shape of a motherly friend here to warm your heart with hot maple syrup. The company, Pinnacle Foods, makes no claim of any historical basis for the character.

What about Mrs. Butterworth's elder competitor **Aunt Jemima:** Fictional. The Aunt Jemima character as a racial stereotype has existed for over 125 years, independent of the food brand, and first made popular in an 1875 minstrel song. Representatives of the Pearl Milling Company, who made the first ready-mix pancake batter, saw an actor playing Aunt Jemima in 1889 (a white male in blackface) and recruited him to represent the product. Only in 1989 was Aunt Jemima's offensively cliché kerchief removed.

Ice cream hippies **Ben & Jerry:** Real. Although you might have confused them with their knockoffs from the 1991 movie *City Slickers* Barry & Ira, Ben Cohen and Jerry Greenfield really did start their own ice cream company and still run it according to their original ideals. They have three separate missions: a Product Mission, an Economic Mission, and a Social Mission. It's worked well enough that they are now in 30 countries.

America's baker **Betty Crocker:** Fictional. Marjorie Husted created and named the character Betty Crocker as an icon for the Washburn Crosby company as it merged with five other milling companies and became General Mills. The name Crocker was an homage to William Crocker, one of the directors of Washburn Crosby. For almost 30 years, Husted went on to provide the voice for the *Betty Crocker Cooking School of the Air* radio program, making the name not just a brand, but a real American icon.

Betty Crocker's challenger **Sara Lee:** Real. In 1951, baker Charles Lubin realized he needed an icon too, and he didn't have one. So, he borrowed the name Sara Lee from his young daughter, who never had anything to do with the company. The Kitchens of Sara Lee grew to $9 million in annual sales before it was acquired by the Consolidated Foods Corporation. The brand grew so successful that the company changed its entire name to that of the brand in 1985, forming the Sara Lee Corporation.

Canned pasta king **Chef Boyardee:** Real. Although he did change the spelling of his name to make his brand more approachable for American consumers, Italian immigrant Ettore Boiardi was indeed a real chef and is indeed the man pictured on the cans. He was the head chef for New York's Plaza Hotel when he left to launch his first restaurant in Cleveland in 1926. Within three years he opened a factory to prepare and ship his spaghetti products nationally. After selling his company to American Home Foods, Chef Boiardi was awarded a gold star from the United States War Department for his efforts producing rations for American soldiers in the Korean War.

Soft drink alchemist **Dr. Pepper:** Fictional. Pharmacist Charles Alderton developed Dr. Pepper's unique taste in 1885, and it was named by his first customer, Morrison's Old Corner Drug Store. A number of stories claim to link Morrison to various doctors named Pepper, but no reliable evidence has ever shown that any of them were the inspiration for the name. In 2009 an antiquer discovered a book containing the formula for a digestive aid called "D Peppers Pepsin Bitters" from Morrison's, so it appears that Dr. Pepper was simply a brand name that the drug store attempted to build.

Dr. Pepper's upstart rival **Mr. Pibb:** Fictional. In 1972 the Coca-Cola Company introduced Mr. Pibb as a knockoff of Dr. Pepper. Coca-Cola has never made any suggestion that there was an actual person named Mr. Pibb. The drink has since been renamed Pibb Extra, and we don't know anyone of that name either.

Cookie cutter **Famous Amos:** Real. Wally Amos was a Hollywood talent agent who sent his own home-made cookies to prospective clients. Although he represented stars such as Diana Ross & the Supremes and Simon & Garfunkle, his cookies were better than his agenting, so in 1975 he opened a cookie store called Famous Amos. Within a few years he was filling orders from supermarkets nationally. But his cookie recipe was better than his cookie marketing, and he had to sell the company only a few years later.

Famous Amos' competitor **Mrs. Fields:** Real. Debbi Fields was only 21 when she opened her first bakery to sell cookies in Palo Alto, CA. The store was so successful that she began franchising, and Mrs. Fields Bakeries has since become one of the amazing American success stories. The company now has more than 650 stores.

Greasy breakfast and barbecue man **Jimmy Dean:** Real. The Jimmy Dean of Jimmy Dean Foods is indeed the same man who sang *Big Bad John* and who portrayed Willard Whyte in *Diamonds are Forever*. He and his brother Don founded the

Jimmy Dean Sausage Company in 1969, and with its popular frontman as a good-humored spokesman, the company did well enough that it was acquired by Consolidated Foods in 1984. By 2004 he was completely retired from the business, and rumored to be hiding out somewhere protected by his security guards Bambi and Thumper.

Pie chef **Marie Callender:** Real. Like some other namesakes, Marie Callender was a real person but had no actual involvement with the company. Her son Don Callender is the one who, in 1948 at the age of 20, opened a wholesale bakery to make pies for the restaurant business, and he named it after her because nobody would want "Don's Pies". The company now operates 139 restaurants throughout the United States, and the frozen food business is owned by ConAgra.

Meat packer **Oscar Mayer:** Real. German immigrant brothers and sausage makers Oscar, Gottfried, and Max Mayer ran a popular meat market in Chicago in 1900. They were among the first meat companies to carry USDA inspection grades beginning in 1906, and were among the first companies to brand meat, first selling it as Oscar Mayer Wieners in 1929. Oscar Mayer was not just a real person; he was one of three real Oscar Mayers to run the company in succession. The first Wienermobile appeared in 1936. The company is now owned by Kraft Foods.

Rice king **Uncle Ben:** Fictional. Like Aunt Jemima and the Rastus character used on Cream of Wheat boxes, Uncle Ben is another in a long line of patronizing and demeaning racial stereotypes associated with foods. Converted Rice Inc. sold rice to the US military during WWII, and owner George Harwell chose the name Uncle Ben in order to appeal to the general public with a fatherly character. Mars, Inc. acquired the company and now claims, almost certainly falsely, that Uncle Ben was simply the name of a successful rice farmer in Texas who was paid $50 to pose for the box photo. They now depict Uncle Ben as the chairman of the board.

So much for food products. Let's go up one level where product names are just a little more important: the worlds of cosmetics and high fashion. It's more important to associate your product with an impressive and fancy sounding name than it is a real name. So let's see if the names they've chosen were chosen because they sound high class, or is it merely that their association with high class products has made the names fancy?

Perfumist **Prince Matchabelli**: Real. I was sure this one was fake, having grown up with the TV commercials of some guy jet setting around Monte Carlo, skydiving, racing speedboats; and yet there never seemed to be any European monarchies missing a prince. Prince Giorgi Machabeli was an amateur chemist and member of the royal family of Georgia. When his Georgian Liberation Committee failed to win independence from the Russian Empire and the Bolshevik Revolution went down, he fled with his wife to the United States in 1921 and launched the Prince Matchabelli Perfume Company.

Cosmetics king **Max Factor**: Real. Watch out for the re-spelled European names. Maximilian Faktorowicz was a young Polish cosmetics expert who worked for the royal family. He emigrated to the United States in 1904 and set up shop in Los Angeles, providing wigs and revolutionary new cosmetics to the growing movie industry. He actually invented the term make-up, and has a star on the Hollywood Walk of Fame.

Outdoor clothier **Eddie Bauer**: Real. Eddie Bauer was a Pacific Northwest outdoorsman who patented the first quilted down jacket in 1940. He made his fortune as a supplier to the US military, and was the first company allowed to use its logo on military issued clothing when he created the B-9 Flight Parka for the US Army Air Corps.

Cosmetics queen **Mary Kay**: Real. Mary Kay Ash worked in sales for over a decade before concluding that women need to run their own businesses instead of being passed over by men. This remains the central marketing theme of Mary Kay Cosmetics, wooing women to join their multilevel marketing scheme hoping to one day earn the iconic pink Cadillac. If Mary Kay's profits from the product you sell covers the cost of the car lease, you owe nothing; if you don't sell enough, you have to pay the lease yourself to keep the car for its two-year term.

Leisure clothier **Tommy Bahama:** Fictional. Come on, have you ever known anyone named Bahama? It's a brand name of Georgia-based Oxford Industries, Inc.

Fashion czar **Tommy Hilfiger:** Real. Tommy Hilfiger is a real fashion designer who took a risk and launched his own brand in 1984, which later went public and gave him a 20-year ride before he sold out to The Man. As a brand, Tommy Hilfiger now sits in the same corner with Tommy Bahama over at Oxford Industries. As a person, Hilfiger jet-sets around, juggling supermodels and reality TV shows and picking fights with Axl Rose in nightclubs.

So if you want to be immortal, name a product after yourself. Just don't do what Thomas Crapper did and become synonymous with bowel movements. Maybe you'll be fortunate enough to have someone do it for you: Oil tycoon Armand Hammer had nothing to do with the baking soda that bears his name. In any case, never underestimate the power of names, and never just assume that a brand or some other icon is real. When it comes to marketing, you should always be skeptical.

References & Further Reading

Amos, W., Robinson, L. *The Famous Amos story: The face that launched a thousand chips.* New York: Doubleday, 1983.

Ash, Mary Kay. *Mary Kay: You Can Have It All: Lifetime Wisdom from America's Foremost Woman Entrepreneur.* Rocklin, CA: Crown Publishing Group, 1995. 258.

Cohen, B., Greenfield, J. *Ben & Jerry's Double Dip: How to Run a Values Led Business and Make Money Too.* New York: Simon & Schuster, 1998.

Ingham, J. *Biographical dictionary of American business leaders; Volume 1 & 2*. Connecticut: Greenwood Press, 1983.

Lee, L. *The name's familiar: Mr. Leotard, Barbie, and Chef Boyardee.* Gretna, Louisiana: Pelican Publishing Company, 1999. 39.

Manring, M. *Slave in A Box: The Strange Career of Aunt Jemima.* Charlottesville: University Press of Virginia, 1998.

42. Organic vs. Conventional Agriculture

Today we're going to take a second look at a pop culture trend that first caught my attention because it so flagrantly waves many of the red flags that characterize pseudoscience: Organic agriculture. I once gave a 15-point checklist of things to look for to help you spot bad science. Organic agriculture is promoted mainly through the mass media, rarely through scientific channels. It's supported by political and cultural campaigns. It relies largely on the "all natural" fallacy. The people promoting it generally have questionable scientific credentials, and they support their claim primarily by pointing out flaws in the norm. These are all characteristic of pseudoscience.

Scientifically, the term "organic food" is meaningless. It's like saying a "human person". All food is organic. All plants and animals are organic. Traditionally, an organic compound is one produced by life processes; chemically, it's any carbon-containing molecule with a carbon-hydrogen bond. Plastic and coal are organic; a diamond is not. So when we refer to organic food in such a way to exclude similar foods that are just as organic chemically, we're outside of any meaningful scientific use of the word, and are using it as a marketing label.

When we try to find common ground, we all agree that healthy food and sustainable production are the goal. So, fundamentally, we're all on the same team, looking for the same thing. Where we split is in our analysis of the history of food production, specifically the role of science in increasing crop yields. Science has brought us crop strains that increase output by factors of 10 and even 20 times even in poor soil, and given us a plethora of tools to combat losses to pests and disease. Generally, science (and the hungry people who benefit) applaud these improvements. Organic proponents (mainly well-fed people) have opposed them, saying they're bad for the environment. To support this position, organic proponents have continued heaping on all sorts of claims about the dangers of modern agriculture: That the food is unhealthy, or that it requires toxic chemicals that poison consumers, ravage the soil, and pollute the oceans with runoff. They poison the well by referring to modern agriculture using weasel words like "chemical farming" and "industrial agriculture". The natural inference we are supposed to make is that organic crops are free from these dangers.

I want to stress that I am not opposed to organic food. It is generally a perfectly fine product. I do have objections to the way it's marketed: It's an identical product, sold at a premium, justified by baseless alarmism about standard food. Whether you agree or not that this alarmism is baseless, you should at least agree that that would be an unethical way to promote a product that offers no real benefit. I choose not to reward this with my food-buying dollar. People who willfully seek out the organic label when buying food are being taken advantage of by marketers employing unethical tactics.

It's a seductive message. Everyone loves to hear that corporations are bad, that all-natural is good, that chemicals and synthetic compounds are poisons. This is not a message that's difficult to sell. It's little wonder that organics have been the fastest growing agricultural market segment over the past decade. It's an ironic little secret that those very same corporate food producers taking our money to sell us organic foods are the same ones spending it on the ad agencies to stoke the anticorporate message that drives them. Nearly 100% of organic food in supermarkets comes from a producer owned by one of the major food companies that also sells regular food. Don't think for a minute that any well-managed food company has not already been on this bandwagon since it started rolling.

I've been pointing out the fallacies of the organic label for some time, so it's frequently assumed that I'm on the payroll of

Big Agriculture. Since Big Agriculture are the ones selling nearly all of the organic food in this country, do you know how stupid that accusation sounds? My motivation is to help people think critically and scientifically, and not simply accept pop culture trends because they sound satisfying, or have been greenwashed with clever marketing.

OK, so I've made some statements about the safety of conventional agriculture, and about the equivalency of organic and conventional produce. It's time to back those up. First, understand that mine is not the extraordinary claim. I'm saying the claims made by organic proponents — that we're all being poisoned — are the extraordinary claims lacking evidence. But since many people seem to prefer the reverse, that the default assumption should be that the entire world population is poisoned with toxic chemicals from conventional agriculture, I'll go ahead and support my "outrageous alarmist claim" that this holocaust doesn't seem to exist.

The biggest misconception is that organic farming does not use fertilizer, herbicides, or pesticides. Of course it does. Fertilizer is essentially chemical nutrient, and the organic version delivers exactly the same chemical load as the synthetic. It has to, otherwise it wouldn't function. All plant fertilizers, organic and synthetic, consist of the same three elements: nitrogen, phosphorus, and potassium. Referring to one as a "chemical" and implying that the other is not, is the worst kind of duplicity, and no intelligent person should tolerate it.

The difference between the two is the source of the chemicals. To make the high-volume commercial versions of both organic and synthetic fertilizer, the source materials are processed in factories and reduced to just the desired chemicals, and the end product, these days, is virtually indistinguishable. Small organic farmers, and home organic farmers, might use fish meal, bone meal, bat guano, or earthworm castings. These are fine products and do indeed deliver the required nutrients. They're just not useful for high volume farming because they're (a) far too expensive, and (b) contain too much ballast, or inactive ingredient, that the crops don't use and merely increase the energy requirements of moving and delivering them.

To make synthetic fertilizer, we start with nitrogen, which we extract from the atmosphere. This process is infinitely sustainable and produces no waste. The potassium is mined from ancient ocean deposits. The phosphorus we get from surface mining of phosphate rock. Although we have centuries of reserves of phosphate rock and millennia of reserves of potassium salts, mining is not sustainable, as these reserves will eventually run out. So, increasingly, producers are turning to seawater extraction for both. This forms a completely sustainable cycle, as the oceans are the ultimate destination of all plant matter and farm runoff.

But clean, sustainable atmospheric and seawater extraction are both taboo for organic certification, which I find astonishing. The chemicals for organic fertilizer must be sourced from post-consumer and animal waste, which is fine but the restriction strikes me as completely arbitrary. Food waste, animal manure, and other organic recyclables collectively provide all the needed ingredients to make refined, high quality fertilizer. The refining process is necessarily a little bit different, but the end product is comparable.

Don't get me wrong — I think fertilizer is a fine use for post-consumer waste. It's certainly better than putting it into a landfill. I'm completely in favor of using all of our restaurant waste, cow manure, or whatever we have in as recyclable and sustainable a way as is practical; and I can't think of a better use than fertilizer. Unfortunately we don't have nearly enough high-quality organic waste to satisfy a meaningful percentage of our food production needs, and developing countries have even less or even none at all; so we're going to need to continue to supplement with sustainably derived synthetics. Why do so many people consider this immoral? I don't know.

Some in the organic lobby have said that organic farming reduces or eliminates the need for added nutrients by rotating crops and better managing the soil. This is true, but it's always been true of all farming, and is in no way unique to organics. Soil management is something all farmers have always done: Describing it as part of the organic process dishonestly implies that the strategy is not also employed by mainstream agriculture. Corn, wheat, and soybeans are the main crops that U.S. farmers rotate. This improves nitrogen content in the soil and reduces the proliferation of pests that thrive on a particular crop. When some farms do practice monoculture, in which only a single crop is grown season after season, it all comes down to a cost-benefit equation. Every farm prefers to avoid the expense of spraying anything they don't have to.

On to pesticides and herbicides. All crops are subject to disease and infestation, and all farmers have to do something about it. Because organic herbicides and pesticides depend on toxic plant-derived chemicals like rotenone and pyrethrin, they've had a tougher time meeting the same standards, making them safe for farm workers and for human consumption, that synthetic versions have already met for decades. Organic versions do meet the standards and are just as safe, but doing so makes them considerably less efficient. According to one winemaker interviewed by the Los Angeles Times, most vineyards do not get certified organic because some of the rules emphasize the ideology over the science. Vineyards need fungicide. Organic fungicide lasts 7 days, while superior synthetic fungicide lasts 21 days. This means two fewer tractors pass through the vineyard spewing diesel exhaust and compacting the soil.

Despite claims in the organic community, there's never yet been a confirmed case of anyone becoming ill from consuming produce contaminated with residue from pesticides or herbicides, either organic or synthetic. Both are certified safe for human exposure, and both are applied at trace levels well below safety standards. In no way does limiting yourself to organic produce decrease your risk of dangerous levels of exposure to pesticides. The risk is practically zero either way.

Everyone has traces of these compounds in their body, no matter what they eat, at ridiculously low parts per billion or even parts per trillion. People who eat conventional produce will usually have safe but detectable levels of conventional pesticides in their body. Organic proponents love to point this out, but somehow they always forget to mention that people who eat only organic produce also end up with safe but detectable levels of organic pesticides in their bodies. If you eat it, it's going to end up in your body, so I'm not sure why this should surprise anyone. Just existing on the planet means that we all naturally have safe but detectable levels of practically every toxic substance imaginable, somewhere in our system. It's too easy to frighten people with such sensationalism. We have to understand the difference between what's safe and normal, and what's harmful.

What harm exists is usually among farm workers who mishandle pesticides, and their plight is among the organic proponents' main arguments against conventional agriculture. Hundreds of thousands of people are diagnosed around the world each year, tens of thousands are hospitalized, thousands die, and it is indeed a real problem that requires increased attention. But there are three very important qualifiers that the organic proponents never seem to mention:

- Two thirds of these cases are deliberate suicides and suicide attempts. In Malaysia it's 73%.
- Illnesses from occupational exposure to organic pesticides are proportional to those from conventional pesticides.
- Almost all of these cases are from developing countries like Indonesia that lack or ignore safety guidelines. Less than 1% of all such illnesses occur in the United States.

The takeaway is that organic practices in no way mitigate such injuries. Buying organic produce does not protect a single farm worker. Following proper safety procedures does protect all farm workers, and this is where resources would be better applied, not in promoting fear about conventional agriculture.

We should choose farming methods that truly address our real concerns — safety and sustainability — not simply methods that satisfy an arbitrary marketing label. To whatever extent these practices include methods that are permitted under organic rules, that's just fine; but there's never a case when a safe, more efficient, and sustainable modern technology that feeds more people worldwide should be disallowed for no logical reason. Buy whatever produce you see in the market that you like and that's cheap, and don't reward the people who are profiteering by selling you fear.

References & Further Reading

Avery, Alex. *The Truth About Organic Foods.* St. Louis: Henderson Communications, L.L.C.; 1ST edition (2006), 2006.

Dangour, A., Aikenhead, A., Hayter, A., Allen, E., Lock, K., Uauy, R. "Comparison of Putative Health Effects of Oragnically and Conventionally Produced Foodstuffs: A Systematic Review." *Food Standards Agency.* Food Standards Agency, 29 Jul. 2009. Web. 12 Jan. 2010.
<http://www.food.gov.uk/multimedia/pdfs/organicreviewreport.pdf>

Hughner, R.S., McDonagh, P., Prothero, A., Schultz II, C.J., Stanton, J. "Who are organic food consumers? A compilation and review of why people purchase organic food." *Journal of Consumer Behavior.* 21 May 2007, Volume 6 Issue 2-3: 94-110.

Kristensen, M., Østergaard, L.F., Halekoh, U., Jørgensen, H., Lauridsen, C., Brandt, K., Bu¨gel, S. "Effect of plant cultivation methods on content of major and trace elements in foodstuffs and retention in rats." *Journal of the Science of Food and Agriculture.* 1 Sep. 2008, volume 88, Number 12: 2161-2172.

MacKerron D.K.L. et al. "Organic farming: science and belief." *Individual articles from the 1998/99 Report.* Scottish Crop Research Institute, 1 Dec. 1999. Web. 22 Jan. 2010.
<http://www.scri.ac.uk/scri/file/individualreports/1999/06ORGFAR.PDF>

Mondelaers K., Aertsens J., Van Huylenbroeck G. "A meta-analysis of the differences in environmental impacts between organic and conventional farming." *British Food Journal.* 1 Nov. 2009, 111, 10: 1098-1119.

43. Should Science Debate Pseudoscience?

Today I'm going to propose a bit of a radical idea. About every week I get invited to debate someone — on another podcast, on the radio, in person — and I'm invited to take the side of science and debate a pseudoscientist. It might be a ghost hunter, it might be a Young Earth Creationist, it might be a practitioner of alternative medicine, always based on some Skeptoid podcast episode I've done that ruffled someone's feathers. Although I used to always accept these invitations, I now always decline them. I have concluded that it is not only useless for science to debate pseudoscience, it is actually counterproductive to science. Today I'm going to argue that no scientist should ever agree to debate a pseudoscientist about a scientific question.

The exception, of course, is court cases, but that's a legal process and not the type of open debate designed to illuminate the public that we're talking about here. In court it's always essential that science put its best foot forward in order to continue strengthening the bond between laws and facts. In court, there are consequences if you lie or make stuff up. In debates, there are no consequences for making stuff up, and there is no tangible benefit for the one who brings the best evidence. Courts of law reward good evidence; debates reward only good rhetoric. A judge is trained to see the difference between evidence and rhetoric, but a debate audience is rarely so well equipped.

I'm not saying that science should not be debated internally. Efforts to falsify existing theory are the core of the scientific method. There is the common analogy of three concentric circles: the center represents the core fundamentals that virtually never change; then the second circle where all the scientific work and research is constantly working out the details, and where scientific debate occurs; then the outer circle is the fringe research which has yet to establish any validity. While scientific debate advances the process of examining and refining the second circle, pseudoscientific debate seeks to throw out the established core fundamentals and replace them with a different set of fundamentals that have never gone through any kind of scientific development. Appropriate scientific debate not only has a place, it's an essential part of the process; but presenting the core fundamentals as if they are comparable to non-scientific alternatives serves no constructive purpose whatsoever.

The primary reason I oppose debates is that a debate, by definition, allows two competing views to be explored and compared, and arguments presented for each. The audience is expected to weigh these arguments and hopefully decide which one they found more compelling. The very nature of a debate presents science as if it is merely a competing opinion. When we agree to a debate, we are agreeing to drag science down to the level of a view that competes with pseudoscience. Simply by agreeing to the debate, we present the scientific method as being vulnerable to disassembly by fallacious pseudoscientific arguments. That's the message we send: Science is not fact, science is merely opinion; and it's as weak as any other.

There's another unfortunate reality about debates, and that's the dirty little not-so-secret that everyone who attends a debate has typically already made up their mind, and has been invited to attend by one side or the other. They are huge proponents of their side, and neither debater has much hope of changing the minds of anyone in the room. Most debates probably have a handful of attendees who are open to actually learning something, but they are an extreme minority. If you've ever attended a debate of any kind, you know what I'm talking about.

When you advertise a debate, maybe 1,000 people will attend. And let's say you do a smashing job and manage to convince that entire handful of convincible attendees that science is real. Great, you won over five people. But what you're forgetting is that for those 1,000 attendees, there are 5,000 people out there who heard about the debate (they saw the ads or flyers or whatever) who did not attend. What you unintentionally communicated to those 5,000 people is that your scientific discipline is academically comparable to the pseudoscientific version, and that both are equally valid. The fact that the debate exists at all struck a blow to the public's perception of the credibility of science that far outweighs any progress you may have made in the room.

I've been the lone representative of science in the room, the one they introduce as "a real trooper for agreeing to come into the lion's den." I've received the condescending smattering of applause from the room where every single person is against me and everything I have to say, but they've "shown me that they're good people too and will treat me respectfully in spite of how misguided I am." Nice folks. And then I'd walk back to my car and every time I'd say to myself "That was a friggin ridiculous waste of time." And I guarantee that their write-up of the event in their newsletter would say I was a nice guy, I was a real trooper to come and talk, and they probably planted within me a seed that would eventually bloom into full-blown science denial, and they'd love to have me back someday to see how that seed has germinated. Going to debate at an event sponsored by the pseudoscience group is always a ridiculous waste of your time. You serve merely as a masturbation enabler for them. Next time, send them a stack of dirty magazines instead.

It has been argued that scientists have a huge advantage in debates because we have the facts on our side. Well, so we do, but that's not an advantage at all. Rather, it's a limitation. The audience members who are not scientists can rarely discriminate between facts and pseudo facts. The pseudoscientist has an unlimited supply of sources and claims and validations. He can say whatever he wants. If compelling rhetoric would benefit from any given argument, he can always make that argument. Pseudosciences have typically been designed around compelling rhetorical arguments. The facts of science, on the other hand, rarely happen to coincide with the best possible logic argument. Having the facts on your side is not an advantage, it's a limitation; and it's a limitation that's very dangerous to the cause of science should you throw it onto the debate floor.

It has also been argued that scientists should debate pseudoscience because if we don't, we allow them to have an unchallenged platform, and the only voice being heard is theirs. I don't buy this argument at all. Not holding a debate doesn't silence us any more than it silences them. Both retain the same "unchallenged platforms" that both have always enjoyed. We have free speech in our society, and anyone who wants to will always have a voice whether we choose to hold a debate or not. What's important is the quality and reach of our voice. I say that science communication should be its own one-way platform. We're the ones who should be refusing to give the pseudoscientists an apparently-equal voice by agreeing to debate them. Science benefits the public; pseudoscience harms the public. We should be doing all we can to promote good science communication, and to refuse to admit the voice of pseudoscience, at every opportunity. They have their free speech already; we don't need to be turbocharging it for them by letting them leech off the credibility we've earned.

I've heard another argument in favor of debating, and that's that when you win, reporters will trumpet that result to a much wider audience. Well, that's an awfully gutsy roll of the dice you're making. Who is this reporter? What makes you so sure he's going to think your position is the stronger? And consider that it's going to be yet another article contributing to the false perception that science and pseudoscience represent two equally valid, debatable perspectives.

I say, take any energies you might be inclined to devote toward preparing for a debate, and instead devote that time to prepare a one-way science presentation that will amaze and enlighten, without any polluted cargo of pseudoscience being delivered alongside. It is not cowardly to protect the delivery of valuable information.

No doubt this chapter will be laughed at by the purveyors of nonsense, saying that I'm advising science to tuck its tail between its legs and run because we know we can't win any debates, because we've discovered that magical thinking is in fact real, and stronger than science. You may well receive this same criticism when you decline a debate. Don't worry about it. Don't feed the trolls. Let the schoolyard bully tease and taunt, you have better things to do than engage him.

Neither am I saying that science should not respond to the promotion of pseudoscience in pop culture. Absolutely we should; my point is that debates are the wrong way to do it. We should write articles like Simon Singh's. We should inject good science into the media through projects like the Science and Entertainment Exchange. We should continue to produce high quality content like podcasts, videos, and TV shows that appeal to a broad audience with entertaining content, and that deliver powerful doses of science education and critical thinking skills; and that never open the door a crack to the contamination of pseudoscience. These are the ways to effectively impact society positively.

So to all of you debaters out there, you may agree with some of my points, or you may disagree with all of them. But like I always say: Whether I'm right or wrong makes no difference. What matters is that you're thinking about these questions. Don't accept any invitation to debate before you consider all of its implications. Science should be taught as fact, not offered as an alternative opinion in a debate.

References & Further Reading

Alters, Brian J., & Gould, Stephen Jay. "Stephen Jay Gould: An Interview." *The American Biology Teacher.* 15 Apr. 1998, Volume 6, Number 4: 272-275.

Beyerstein, B. "Distinguishing Science from Pseudoscience." *The Center for Curriculum and Professional Development.* 1 Jul. 1995, October 1996.

Dawkins, Richard. "Why I Won't Debate Creationists." *RichardDawkins.net.* Upper Branch, 15 May 2006. Web. 17 Aug. 2009. <http://richarddawkins.net/articles/119>

Krauss, Lawrence. "Odds Are Stacked When Science Tries to Debate Pseudoscience." *New York Times.* 30 Apr. 2002, F: 3.

Lilienfeld, S., Landfield, K. "Science and Pseudoscience in Law Enforcement: A User-Friendly Primer." *Criminal Justice and Behavior.* 1 Oct. 2008, Volume 35, Number 10: 1215-1230.

Novella, Steven. "Simon Singh's Libel Suit." *SkepticBlog.* Skeptic Magazine, 11 May 2009. Web. 17 Aug. 2009. <http://skepticblog.org/2009/05/11/simon-singhs-libel-suit/>

44. Decrypting the Mormon Book of Abraham

Today we're going to point our skeptical eye at one of the supposedly ancient scriptures of the Mormon Church, the *Book of Abraham*. In 1835 the Church came into possession of some Egyptian papyri, said to have been translated with divine guidance by their prophet, Joseph Smith. Smith reported that the papyri were "the writings of Abraham while he was in Egypt, called the *Book of Abraham*, written by his own hand, upon papyrus." In the book, Jehovah reveals to Abraham the nature of the universe and the order of all things, in a personal conversation, including knowledge of the planet Kolob, which is close to where God lives.

Sometime in the early 1800's, an antiquities dealer named Antonio Lebolo returned from Egypt with eleven mummies and other artifacts, including papyri, from the region around Thebes. Upon his death, the collection was sold at auction, and ended up in an exhibition that traveled the United States, which sold the artifacts off as it went. In 1835 this exhibition reached Kirtland, Ohio, the headquarters of the Latter-Day Saints. The exhibition's proprietor at the time was a Michael Chandler, who was well aware that the church had been founded upon Joseph Smith's claimed translation of gold plates written in Egyptian, which became the *Book of Mormon*. Chandler gave Joseph Smith a viewing of the collection, which by that time had been reduced to four mummies and a few rolls of papyri containing hieroglyphics, and Smith gave Chandler a cursory translation of some of the papyri.

Shortly thereafter, two church elders, Joseph Coe and Simeon Andrews, purchased the entire collection from Chandler for $2,400, about $60,000 in today's dollars. Smith then took the papyri into seclusion to translate them. At his side were Oliver Cowdery and William Phelps who transcribed. The product of their labors is the *Book of Abraham*. It's not very long; five short chapters, less than six thousand words. The book includes three Egyptian-looking illustrations done by Reuben Hedlock, a professional engraver who copied them from the actual papyri. The *Book of Abraham*, with its illustrations, is now included in the *Pearl of Great Price*, one of the Mormon church's five books of scripture.

Upon Joseph Smith's assassination in 1844, the artifacts were passed to his mother, and then to his widow, who sold them to a collector by the name of Abel Combs. Combs broke up the collection, and about half the artifacts went to the Wood Museum in Chicago, where they were subsequently lost in the Great Chicago Fire of 1871. The whereabouts of Joseph Smith's papyri remained a mystery for nearly a century, until a scholar named Dr. Aziz Atiya from the University of Utah happened upon them in the New York Metropolitan Museum of Art's archives in 1966, recognizing them by one of the illustrations that he knew from the *Pearl of Great Price*. Upon investigation, it was discovered that the Metropolitan had purchased them in 1908 from the daughter of Abel Combs' housekeeper, including an affidavit from Smith's widow. All of Smith's original papyri had been fragmentary, and these ten pieces probably made up some one-third to one-half of his original collection. The Church bought the papyri from the Metropolitan and brought them back to the Salt Lake City headquarters, where one additional fragment was discovered in the Church's own archives; bringing the total count of Joseph Smith's original papyri that survive today to eleven.

Having the original documents available made it possible for Egyptologists to examine and properly translate them, to see whether they do indeed match what Smith, Cowdery, and Phelps came up with. If they were indeed divinely inspired with translating abilities, you'd think that would be the case. Let's find out.

This is a good time to introduce Thomas Stuart Ferguson, an attorney, amateur archaeologist, author, and Latter Day Saint. Ferguson's lifelong passion was finding archaeological evidence from Mesoamerica that confirmed the *Book of Mormon* stories. His book *One Fold and One Shepherd* is considered one of the seminal works on the subject. It was Ferguson who first approached Brigham Young University and persuaded them to create a Department of Archaeology. He founded the New World Archaeological Foundation to bankroll expeditions to Mesoamerica, and even got the Church itself to become a major sponsor of his work.

So imagine Ferguson's excitement at the opportunity to provide a real live black-and-white proof of Joseph Smith's divine inspiration, and an actual historical document, thousands of years old, telling the *Book of Mormon* stories. Ferguson obtained photographs of the eleven papyrus fragments and sent them to Klaus Baer, a professor of Egyptology at the University of Chicago, and to an unaccredited amateur, D.J. Nelson. He also sent copies to a pair of Egyptologists at U.C. Berkeley, Professor Henry Lutz and Leonard H. Lesko, but provided no information about their origin or anything that might link them to the *Book of Abraham*. All four men quickly came back with the exact same proper identification of the documents.

They were examples of what's called a hypocephalus, meaning "below the head". This is a round papyrus or other inscribed object placed under the head of a deceased person for burial. No two are the same. They are inscribed with a traditional funerary text, often from *The Book of the Dead*, and this particular one was *The Breathing Permit of Hôr*. The papyri were merely unremarkable burial trappings, quite likely from Antonio Lebolo's original mummies. They had nothing remotely to do with Abraham, the planet Kolob, or anything else found in Joseph Smith's translation. Moreover, numerous Egyptologists since have examined the widely published photographs, and identified in detail everything found in the illustrations. Again, Smith's own callouts and identifications bear no resemblance to the actual contents. Ferguson said "I must conclude that Joseph Smith had not the remotest skill in things Egyptian-hieroglyphics."

The Church has defended Smith's claim against the findings of academia. Hugh Nibley, a late professor of Mormon scripture at Brigham Young University, was the Church's primary apologist for many years. Nibley's main defense was that the papyrus fragments recovered from the Metropolitan did not happen to be the same ones in which Smith found the *Book of Abraham*, and thus the different translations; after all, perhaps as much as two-thirds of the original papyri have never been recovered. Ferguson scoffed at this suggestion, pointing out that all three of Reuben Hedlock's illustrations exactly match those in the existing papyri.

Smith, Cowdery, and Phelps had also written the *Egyptian Alphabet & Grammar*, purportedly a guide for understanding the heiroglyphs in the documents they translated, which has remained in the Church's possession. It makes clear references to the heiroglyphs and their positions on the pages, unambiguously referring to the existing papyri. They are clear, additional evidence that the existing papyri are the ones claimed to contain the *Book of Abraham*. Nibley dismissed the *Egyptian Alphabet & Grammar* as "of no practical value whatever and never employed in any translation." I have to agree with Nibley here: They certainly do not seem to be of any practical value, but that says nothing about the finding that they do reference the existing papyri.

The papyri have been dated to the first century BC, about 1500 years after Abraham is claimed to have lived, which makes it difficult to reconcile Smith's statement that they were written by Abraham's own hand. Hugh Nibley came to the Church's rescue again, stating that it's common to refer to a book as having been written by someone without literally meaning that that exact volume was created by a pen held in that person's own hand. The Church itself goes even farther, stating that "Joseph Smith never claimed that the papyri were autographic (written by Abraham himself)", implying that there's still a loophole for Smith's claims to be true.

But this is a tenuous position to which to cling. Joseph Smith's introduction to the *Book of Abraham* reads:

> *The writings of Abraham while he was in Egypt, called the Book of Abraham, written by his own hand, upon papyrus.*

Joseph Smith also showed a papyrus to Charles Adams, the son of John Quincy Adams, who reported that Smith told him:

> *This...was written by the hand of Abraham and means so and so. If anyone denies it, let him prove the contrary. I say it.*

Beyond any reasonable doubt, Joseph Smith maintained that his papyri were literally written by Abraham's own hand, and that they told Abraham's story. Both are, beyond any reasonable doubt, untrue.

It's not possible to get inside the heads of Joseph Smith, Oliver Cowdery, and William Phelps, so we can't really know what their honest intentions were. The most cynical analysis concludes that the *Book of Abraham's* authorship was a fully deliberate fraud, where all three men knowingly conspired to contrive a Bible-style book to add to their doctrine, claiming the papyri as the source when they well knew that it probably had nothing to do with the story they invented. A more charitable version of events has Smith honestly believing he was divinely inspired to translate the papyri, reeling off the tale as it came to him, with Cowdery and Phelps sincere in their faith and transcribing Smith to the best of their ability. Maybe Smith alone knew he couldn't read the papyri and was making up the story, and hoaxed Cowdery and Phelps. All we can say for sure is that its source is absolutely not what the Church claims it is.

The Church says that the significance of the *Book of Abraham* is that it is "evidence of the inspired calling of the Prophet Joseph Smith." I can find no rational argument that supports this. It is merely evidence that the talents of Smith, Cowdery, and Phelps, combined with any divine inspiration any of them may have had, were insufficient to translate a document that is a trivial task for any knowledgeable Egyptologist. Honest Mormons should have grave concerns over the Church's continued promotion of a claim proven to be false. It's time for Mormons with intellectual integrity to demand the *Book of Abraham* be reclassified as not based on any Egyptian papyri, and its authorship properly assigned to Smith, Cowdery, and Phelps, with whatever status the Church likes that does not endorse the bogus translation.

REFERENCES & FURTHER READING

Adams, C. F. *Diary of Charles Francis Adams.* Boston: Massachusetts Historical Society, 1844.

Larson, Charles M. *By his own hand upon papyrus: A new look at the Joseph Smith papyri.* Grand Rapids: Institute for Religious Research, 1992.

LDS Church. "The Book of Abraham." *Institutes of Religion.* The Church of Jesus Christ of Latter Day Saints, 1 Jan. 2008. Web. 12 Jan. 2010. <http://institute.lds.org/manuals/pearl-of-great-price-student-manual/pgp-3-a.asp>

Parker, R. "The Joseph Smith Egyptian Papyri - Translations and Interpretations." *Dialogue: A Journal of Mormon Thought.* 1 Jul. 1968, Volume 3, Number 2: 86.

Ritner, Robert K. ""The Breathing Permit of Hôr" Among The Joseph Smith Papyri." *Journal of Near Eastern Studies.* 1 Jan. 2003, Volume 62, Number 3: 161-180.

Webb, R. C. *Joseph Smith of the Church of Jesus Christ of Latter Day Saints as a Translator.* Whitefish, MT: Kessinger Publishing, 2004. 3-18.

45. SHOULD TIBET BE FREE?

Perhaps an equally important question is "Should a science journalist take on a political topic?" For a long time, people have been asking me to research the subject of Tibet, and for a long time I've been putting the requests into a folder and keeping it stored away. My show is called Skeptoid, not Politicaloid, and my purpose is not to advocate one side or the other in political questions where you have two sides that are perfectly valid to different groups of people. But the more requests I've received, the more I've realized that there is a lot of misinformation, if not true pseudoscience, surrounding Tibet. There is, undoubtedly, a set of popular pop-culture beliefs out there, based entirely upon made-up crap that bears little resemblance to reality.

Mind you, I'm not saying "Hey, you've heard one side, let me give you the other side," because that's the job of a political commentator. What I'm saying today is "Here is the reality of Tibet, go forth and form whatever opinion you like," but base it on reality, not on made-up metaphysical nonsense. I'm encouraging you to apply skepticism to the reasons you may have heard for freeing Tibet.

Like most Americans, I grew up watching video of the Chinese army taking howitzers and destroying the massive centuries-old Tibetan monasteries in 1959, and that's an indisputable crime against history, religious freedom, and the dignity of Tibetans. And then I watched video of the Dalai Lama, the exiled spiritual and political leader of the Tibetan people, in his red and yellow robes, speaking words of wisdom and brotherhood and freedom and peace. And I'll freely admit: For nearly all of my life, this was the extent of my knowledge of the Tibet situation: Violence and cruelty from the Chinese; innocence and beauty from the Tibetans. I believe that many Westerners, including many who fervently wave *Free Tibet* placards, have little knowledge of the situation any deeper than that. But isn't it likely that there's more to it than that? Isn't it equally disrespectful of the Tibetans as it is of the Chinese to attempt to encompass who and what they are with those tiny little pictures?

A complete history lesson is impossible, but here's a quick overview of the points relevant to today's discussion. China and Tibet have a long and complicated history. In 1950, China invaded to assert its claim, and ruled by trying to win hearts and minds, building roads and public utilities, and allowing the Tibetan system of feudal serfdoms to remain largely intact. In 1959 the Tibetan ruling class revolted, prompting a Chinese crackdown that sent the Dalai Lama and most other Tibetan aristocrats into exile in India, where they remain to this day. The former serfs became ordinary Chinese citizens, and Tibet is now an "autonomous region" in China, a status that many describe as actually less autonomous than an ordinary Chinese province. From his palace in India, the Dalai Lama now travels the world in a private jet, hobnobbing with the wealthy and powerful, fundraising, and writing highly successful books on metaphysics.

Recently there were some anti-China, pro-Tibet protests in Nepal, a neighboring independent nation. This is illegal in Nepal, and the authorities have been cracking down on it. Why does Nepal side with China on this issue? Because they depend heavily on Chinese aid to survive, and this is a requirement that China imposes, though they call it a "request". At first glance you might be shocked that an independent nation would give up its freedom of speech to make a deal with the devil, but that's an easy opinion to form when you're not hungry. It makes sense for Nepal to agree to these terms, because their back is against the wall: They need China's aid. As for China imposing this condition? Well, that's one for you to chalk up in your column of "Things China Needs to Reconsider".

So, why doesn't China simply give Tibet the same treatment they give Nepal — let them be an independent nation, give them aid, and just require them to say only nice things about them? Well, Nepal has long been an independent nation; Tibet hasn't. The history of China's rule over Tibet is exceptionally complicated and goes back many centuries. Anyone who tells you that either Tibet is historically part of China, or that Tibet is historically free, is making a disingenuous oversimplification. Personally, I choose to discount this subject completely, and not because it's too intricate to make a clear decision. I discount it because practically every square inch of land on the planet has been taken over militarily or annexed or stolen in one way or another from one people by another people. We don't give California back to the Spanish, and we don't give Italy (once sacked by Vikings) back to Norway. Ancient history is not the way to settle current border disputes. To find a meaningful settlement that makes sense for people today, you have to consider Tibet to be a current border dispute. So while we're chalking up China's claim of ancient possession in the column of "Things China Needs to Reconsider", let's also chalk up Tibet's claim of ancient autonomy in the column of "Things Tibet Needs to Shut Up About".

And once we open up that column, we find it's a Pandora's Box. Advocates of a free Tibet make a long list of charges against Chinese oppression, largely centered upon a loss of rights and freedom. This claim makes anyone familiar with Tibetan history cough up their coffee. The only people who lost any rights under Chinese rule are Tibet's former ruling class, themselves guilty of cruelty and oppression of a magnitude that not even China can conceive. The vast majority of Tibetans, some 90% of whom were serfs, have enjoyed a relative level of freedom unheard of in their culture. Until 1950 when the Chinese put a stop to it, 90% of Tibetans had no rights at all. They were freely traded and sold. They were subject to the worst type of punishments from their lords, including gouging out of eyes; cutting off hands, feet, tongues, noses, or lips; and a dozen horrible forms of execution. There was no such concept as legal recourse; the landowning monk class *was* the law. There was no such thing as education, medical care, sanitation, or public utilities. Young boys were frequently and freely taken from families to endure lifelong servitude, including rape, in the monasteries. Amid all the pop-culture cries about Chinese oppression, why is there never any mention of the institutionalized daily oppression levied by the Dalai Lama's class prior to 1959?

Free Tibet advocates also point to the destruction of Tibetan culture. This charge is particularly bizarre. The only art produced in Tibet prior to 1950 was limited to the output of a few monks in each monastery, principally drawings of monasteries. New literature had not been produced in Tibet for centuries. Since the 1959 uprising, art and literature in Tibet have both flourished, now that the entire population is at liberty to produce. Tibet even has its share of well known poets, authors, and internationally known artists now.

Make no mistake about China's history of human rights failings: China's "Great Leap Forward" and "Cultural Revolution" programs from 1958 through 1976 were as disastrous for Tibet as they were for the rest of China. There can never be any excuse for the deliberate widescale destruction of life, liberty, and property during those years. Hundreds of thousands of Tibetans, and tens of millions of Chinese, lost their lives during this misguided pretense at "reform". This was a phase that China went through, and it's arguable that Tibet would have been spared this torment if they had been independent at the time. But for your average Tibetan in the field, a serf with no rights, living and working and dying at the whim of his lord, were those decades really worse than they would have been without China? There's no way to know, but to a skeptical mind, it's not a slam-dunk that China's Cultural Revolution was harder on Tibet than Tibet's ruling class had always been in the past.

If we think back to our list of red flags to identify misinformation, cultural campaigns and celebrity endorsements should always trigger your skeptical radar. Few campaigns are as near and dear to the hearts of Hollywood activists as "Freeing Tibet". Notable Tibet advocates include Sharon Stone, Richard Gere, Paris Hilton, and the great political science scholar Lindsay Lohan. Journalist Christopher Hitchens notes that "when on his trips to Hollywood fundraisers, [the Dalai Lama] anoints major donors like Steven Segal and Richard Gere as holy." Being anointed as holy probably does great things for your social standing within Hollywood, but it should not be considered evidence of expertise. I'll bet that if you asked either Steven Segal or Paris Hilton to lecture on the events of the Lhasa Uprising of 1959, you'd find that neither knows even the most basic information about the cause they so passionately advocate. Just because Hollywood celebrities promote a viewpoint doesn't mean they're qualified to do so, something that (unbelievably) still seems to escape most people.

Furthermore, the people shouting loudest about freeing Tibet don't seem to be aware that that's not even what the Dalai Lama wants from China. He's not seeking full independence, Nepal style; rather he would like to achieve the same status as Hong Kong, which is a "special administrative region". This would give them full economic benefits without having to become a regular province, something along the lines of a US territory. So here's a note to all the Hollywood celebrities: If you really want to support the Dalai Lama, ditch your "Free Tibet" signs and paint some up that say "Change Tibet from an Autonomous Region to a Special Administrative Region". It's not as good of a sound bite, and it's a change that would have little practical impact on Tibetans; but it would allow the Dalai Lama to return to his aristocratic lifestyle and his 1,000 room palace at Potala.

So to all those who so heatedly call for the freeing of Tibet, first consider whether you have the expertise to know whether Tibet is best served as an autonomous region or as a special administrative region. Understand exactly what implications such a change may have upon the economics and the daily lives of its citizens, or maybe even entertain the possibility that it's a decision best left to Tibetans.

References & Further Reading

Beckwith, C. *The Tibetan Empire in Central Asia.* Princeton: Princeton University Press, 1987.

Goldstein, M. *The Snow Lion and the Dragon: China, Tibet and the Dalai Lama.* Los Angeles: University of California Press, 1995.

Grunfeld, A. *The Making of Modern Tibet, revised edition.* Armonk: M. E. Sharpe, Inc., 1996.

Parenti, M. "Friendly Feudalism: The Tibet Myth." *Political Archive.* Michael Parenti, 1 Jan. 2007. Web. 14 Dec. 2009. <http://www.michaelparenti.org/Tibet.html>

Sperling, E. *The Tibet-China Conflict: History and Polemics.* Washington: East-West Center, 2004.

46. How to Be a Skeptic and Still Have Friends

When I first started the Skeptoid podcast, it was anonymous. I didn't give my name at all in the first 5 or 10 episodes. Why? Because I was afraid of offending my friends and family, afraid of becoming known as a hateful, closed-minded skeptic. I knew that skeptical outreach is an important educational task and needs to be done, but at the same time, I had to live with people. I have friends, neighbors, coworkers, and acquaintances. I'd long ago learned to keep my mouth shut when someone started praising the merits of some new paranormal thing. I reasoned that the best way to be a skeptic *and* a member of society was to broadcast my critical thinking analyses as far and wide as I could with the podcast, while shielding myself inside a safe little bubble of personal anonymity.

As you probably know, that didn't remain the case. My friends and family found out about Skeptoid almost immediately, but since they're great people I didn't really catch much flak for it, and they pretty much knew that I was skeptical anyway. I decided at that point that if I was going to do it at all, I was going to do it all the way. So I re-recorded all the early episodes with my name on them, and put my full name and picture and bio on the web site. In for a penny, in for a pound; so the saying goes.

This doesn't mean that I go out looking for fights. I rarely go out of my way to proactively challenge a friend's belief when I happen to overhear them talking about something paranormal or pseudoscientific, and then only when it's appropriate to do so. Does it irk me when I hear them talking about how scary the ghosts were in last night's episode of *Ghost Hunters*, or how they're treating their back pain with reiki? Absolutely it does.

And now, since they know that I'm "that skeptical guy", frequently they'll come up to me with something. More often than not, they saw on Action News last night that you can run your car on water, or that an old man somewhere is a proven psychic healer, or that an engineer has gone on record saying the Twin Towers were a controlled implosion. Usually they're snickering because they often believe that *now* they've got me, that *now* their evidence is irrefutable and they're about to go one-up on the skeptic. They ask me what I think.

And I'm not the only one who gets this. I frequently hear from people who find themselves in similar situations. Here's one email I got:

> *In my experience, I always come off as a "know-it-all jerk" because in conversations, I have ethical issues with just letting them go on in life with misinformation, especially as a scientist. Maybe the real lesson is just "don't talk science with them. let them believe what they want to believe." I just feel so terrible knowing that they're going to go on, and maybe make a poor decision based upon the assumption that what they know is true.*

And here's another:

> *Starting an argument, although productive, is not instrumental in making and keeping friendships, since a lot of people are very happy with their delusions and only become annoyed with someone when their false knowledge is pointed out. How does one go about informing a true believer without alienating them?*

So there we have it. The problem: How to be a skeptic, and talk with your friends and coworkers when the subject comes up, without turning people off? Here is the solution.

Focus on where you agree, never on where you disagree. *Start by finding common ground.* No matter who you're talking to, they have *some* level of skepticism about *something*. Ask them, "Isn't there some myth you've heard that you don't necessarily believe?"

They'll answer "Well sure, Bigfoot, space aliens," whatever.

Tell them "I'm skeptical of Bigfoot for the exact same reasons you are. Tell me why you don't believe in Bigfoot?" And now you've got your friend *telling you* the very reasons you're skeptical of the new claim. The evidence is of poor quality, it's too improbable, whatever it is. Help your friend along. Point out more reasons to be skeptical of Bigfoot. Be familiar with our checklist of 15 warning signs to help you spot pseudoscience given in episode 37 of the Skeptoid podcast.

And then, once you have a good list, apply that same reasoning to the new claim. "We agree that part of the reason Bigfoot is suspect is that we have low quality evidence coming from people with dubious credentials. We can also find those same problems with the claim that you can run your car on water. Also, we agree that one reason Bigfoot is improbable is that if it was real, we'd have known about it by now — people have been living in Bigfoot habitat for hundreds of years. We can say that same thing about running your car on water — science has known all about oxyhydrogen and electrolysis for hundreds of years and exploited it many different ways. It wouldn't have to wait for some guy on the Internet to claim to know something that science doesn't."

Feelings are hurt not so much when there is disagreement, but when someone is summarily proclaimed to be wrong. This doesn't just mean telling your friend that he's wrong; it includes telling your friend that the TV Action News is wrong. It's your friend's source, he found it convincing; and when you simply declare it to be wrong without having seen it, you come off as petty and dismissive. Avoid negative language. Avoid saying that anyone is wrong.

"That announcement you heard alerts us to the possibility that this new breakthrough is true, just as the recent Bigfoot news story alerted us that a Bigfoot body might have been found. But science doesn't determine the validity of a theory based on whether or not its proponents have sent out a press release; an announcement that would be *more* interesting would be that the test protocols have been published and the experiment is being successfully replicated all over the world."

Find the common ground. "What are you skeptical of? Well, I'm probably skeptical of it for the exact same reason you are."

You can also accept your friend's claim as a great first step, and tell him what else you'd need to see to be convinced. When something's real, it's real; it can be defined, measured, quantified, and replicated by other researchers. When something's only published on the fringe, or reported from only a single source, that doesn't make it wrong yet; but it does mean that it has not yet been replicated by objective scientists following the same protocols.

Just because you *can* have this conversation in a positive and non-adversarial way doesn't mean you always have to have the conversation. I still find it best to simply keep my mouth shut a lot of the time. A neighbor knocked on my door with his acupuncturist's business card when I was suffering from some pain after one of my volleyball surgeries. That's a kindness, and I thanked him and left it at that. This way the neighbor remains my friend and the door is always open to have the conversation at a more appropriate time.

Spreading critical thinking by engaging in conversation with your acquaintances should be a way to build bridges, not to expose rifts. If you take one thing away from this, it should be that point. Concentrate on where you agree. I've found that this has converted people who used to come to me as an adversary to challenge me with new claims into friends who seek out my opinion on stories that sound fishy to them.

The important first step is to allow yourself to become known as a skeptic. Wear the T-shirts and have the books sitting on your desk at the office. When people know that you're the skeptic, they'll *come to you* when they want to challenge you. And when they come to you, you're not the jerk. Put yourself out there as a skeptic, and wait for the business to come to your door. When it does, handle it positively and show people that they're skeptics too, they just didn't realize it. When you can help someone to understand that they are already themselves a skeptic of something — Bigfoot, aliens, UFOs, celebrity psychics, whatever — your job is half done. It's like judo, use your opponent's strength against him. Help him redirect his own intelligence and existing skepticism towards the subjects where he has not yet thought critically. In this way, you can be a skeptic and still have friends — and, chances are, you'll even have new skeptical friends.

References & Further Reading

Burgess, G., Burgess, H. "Crafting Effective Persuasive Arguments." *Conflict Information Consortium*. University of Colorado, 30 Dec. 1998. Web. 25 Aug. 2008. <http://www.colorado.edu/conflict/peace/treatment/usepersn.htm>

Goldstein, N., Martin, S., Cialdini, R. *Yes! 50 Secrets from the Science of Persuasion*. London: Profile Books, 2007.

Novella, S. "How to Argue." *The New England Skeptical Society*. The New England Skeptical Society, 1 Mar. 2009. Web. 23 Jan. 2010. <http://www.theness.com/how-to-argue/>

Sagan, C., Druyan, A. *The Demon-Haunted World: Science as a Candle in the Dark*. New York: Random House, 1995.

Wilson, R. *Don't Get Fooled Again - The Skeptic's Guide to Life*. London: Icon Books, 2008.

47. Daylight Saving Time Myths

Today we're going to turn our clock back (or forward, as the case may be) and screw up our sleep cycles, casting everything into disarray for a good week until we recover — for today's topic is daylight saving time: The myths, the fallacies, and the facts. Why on Earth would we ever want to change our clocks a few times a year? Is it actually a good idea? In today's day and age, is it still good, or is it outdated?

Benjamin Franklin is often credited with inventing the concept of daylight saving time, on the principle of conserving candles. However, a closer look at this popular tale reveals cause for skepticism. Franklin's paper was actually a satire poking fun at the partying lifestyle of Parisians, not a serious recommendation. It was a letter written in 1784 for the *Journal de Paris* in which he proposed firing cannons at sunrise to wake people up and break them from their nocturnal habits. To justify this, he suggested that sleeping instead of burning candles all night long would save Parisians 64 million pounds of candle wax over six months.

The first serious propositions came independently from different sides of the world a century later: from New Zealander George Hudson in 1895, and Englishman William Wennett in 1907. Both recommendations were to provide for additional sunlit leisure time after work, and it was proposed for use in the summertime only because if it were done in winter, the shorter days would force morning activities (like children walking to school) to happen while it was still dark. That's why we have the current schedule of observing daylight saving only in the summertime: Our workday squeezes into the middle of that narrow band of daylight during the winter, and it drops down to the bottom of the wider band of daylight in the summer; keeping our morning wakeup times roughly aligned with sunrise, but giving us an abundance of extra daylit playtime after work when the days are long enough to permit.

Perhaps the most pervasive popular perception about daylight saving time is that it's all about farmers, the idea being that certain farm tasks should be done at sunrise, whether it's milking cows or watering or harvesting crops, and changing the clock makes this easier somehow. The obvious response to this is that these tasks are going to continue to be done at sunrise, regardless of the time shown on some irrelevant clock. When you dig in and read the arguments for or against daylight saving put forth by various groups, farmers are said to be among the most vocal opponents of daylight saving. Here's a quote that I must have found a hundred times in different sources, word for word:

> *Farmers, who must wake with the sun no matter what time their clock says, are greatly inconvenienced by having to change their schedule in order to sell their crops to people who observe daylight saving time.*

No matter how many times I found that same sentence, I could never discover its original source. The concept is illogical at face value. In the morning hours, daylight saving's effect is to keep the clock *more in line* with sunrise: i.e., 6:00am comes an hour earlier in the summer when the sun is rising earlier. If farmers need workers to arrive when the sun starts drying the dew, daylight saving is clearly their friend. I found many, many articles repeating the presumption that farmers oppose daylight saving, but almost nowhere did I find a good reason articulated; at least not one that pertains to farming. The president of the New South Wales Farmers Association could only come up with this argument:

> *Daylight saving has a significant adverse impact on rural families and communities businesses. An example is children traveling home from school in the heat of early afternoon sun. Members also say they have trouble getting their children to*

bed at normal bed times, and farmers indicate that they work longer days.

...which says nothing about daylight saving creating a problem for the practice of farming. At the same time, other farmers express the same pleasure as other people at an extra hour of sunlit family recreation time after a summer workday.

Dairy farmers have the closest thing to a cogent argument that I could find. Let's say that their product has to be to market at a given clock time, and twice a year the clock changes by an hour. This forces the cows to be milked at a 23 or 25 hour interval, once each year, instead the 24 hour interval to which they're accustomed. Evidently this disruption to the cow's schedule is problematic for the cow. If this really is a significant problem for cows, it constitutes one of only two farm-related arguments against daylight saving that I could dig up.

The other is not unique to farms, and has to do with moving heavy equipment on roads in the early morning. Many such vehicles can only be driven during daylight, and many of the rest shouldn't be operated in darkness for safety reasons. When daylight saving drops the working man's start time down closer to sunrise during the summer, this problem is prolonged for more of the year. But I'm still not convinced by this argument's logic. Farm work starts with the sun. Safe equipment operation starts with the sun. That's how it's always going to be. Daylight saving keeps the clock time at the beginning of the workday more consistent with sunrise year round, so, from a rational perspective, the people who depend on the morning sun for their job have every reason to be the biggest supporters of daylight saving: It brings them better consistency.

So now we come to what everyone believes, and what's written down on paper as the official reason we observe daylight saving time: The conservation of electricity. The idea is that residential power usage is reduced because people don't have to turn their lights on until an hour later in the summer. This was indeed true the first time the question was deeply studied, which was during the oil crisis of the 1970's. The Department of Transportation calculated a 100,000-barrel savings in oil, from a 1% savings in power usage, compared to if we'd stayed on standard time. However, in the decades since then, air conditioning has become much more ubiquitous, and its power consumption greatly outpaces the reduction in lighting — although this is slightly offset by reduced heating on fall and spring mornings. In addition, people have many more electronic gizmos around the house then they did in the 1970's. Having people at home for an extra hour is not nearly such a great way to conserve electricity as it used to be.

Nowadays, people studying power usage during daylight saving get mixed results. There's a lot of regional variation. Places like Florida, with maximum air conditioning needs, clearly use more electricity because of daylight saving; while cooler northern states may still see overall savings due to reduced lighting needs. Generally, reports of national energy consumption combine results from a cross section of utility companies nationwide. Depending on what utilities you include in your report, you may get very different overall results. A truly comprehensive report that accounts for all data is probably outside the realm of practicality.

If you do a Google search for daylight saving energy consumption, you'll find reports are all over the map. In fact, the year after the Department of Transportation found a 1% savings, the National Bureau of Standards reviewed their data and found *no* savings. Some reports find that it uses as much as 1-2% additional electricity; some find there's a savings of as much as 1-2%. Most fall within the statistical margin of error. The one statement I'd feel comfortable making with authority is that any possible energy savings that may be derived from daylight saving time is statistically insignificant.

There's one powerful reason that daylight saving is probably here to stay, and it has nothing to do with farms or electricity or road safety. Strong reasons usually have to do with money. Not money that you send to your utility company, but money that you hand over at the cash register. During the warm summer months when it's possible to do so in comfort, people like to be out and about in the evening. They like to go out for dinner, drinks, or a movie, or wander through stores and galleries. They also like to play golf and tennis. Whenever they do these things, they spend money. Lots of money, in the collective. Give them an extra hour to recreate in the summer, and they spend even more money. In 1986, an extra month of daylight saving was added to the calendar, and representatives of various recreational industries appeared in front of Congress to testify about the effect it had on their bottom lines. The golf industry is said to have benefited by an additional $200 million, just from that one additional month; and the barbecue industry is said to have sold an additional $100 million in barbecues and charcoal briquettes. The additional extension in 2007 into November was supported strongly by the candy industry, who can sell a lot more Halloween candy when kids can spend an extra hour trick or treating before bedtime.

There is even a certain lobby out there that points to the environmental impact of this extra hour of shopping, dining, and golf. One paper from UC Santa Barbara calculated the cost of the resulting pollution as several million dollars per year. Whatever your particular fancy, you can probably find someone who's written a paper saying that daylight saving is good or bad for it. Daylight saving is one case where the fewer words you use to describe it, the more accurate you are. One word: Money. The more details you go into beyond that, the more treacherous your footing.

References & Further Reading

AAP. "NSW: Farmers work longer and kids won't go to sleep: association." *AAP General News Wire.* 22 Oct. 2007, Wire: 1.

Aries, M.B.C., Newsham, G.R. "Effect of daylight saving time on lighting energy use: A literature review." *Energy Policy.* 1 Jun. 2008, Volume 36, Issue 6: 1858-1866.

Downing, Michael. *Spring Forward: The Annual Madness of Daylight Saving Time.* Berkeley: Shoemaker & Hoard, 2005. 147-151.

Franklin, B., Goodman, N. *The ingenious Dr. Franklin: selected scientific letters of Benjamin Franklin.* Philadelphia: University of Pennsylvania Press, 1931. 17-22.

Kotchen, M.J., Grant, L.E. "Does Daylight Saving Time Save Energy? Evidence from a Natural Experiment in Indiana." *NBER Working Paper Series.* National Bureau of Economic Research, 1 Oct. 2008. Web. 22 Sep. 2009.
<http://www2.bren.ucsb.edu/~kotchen/links/DSTpaper.pdf>

48. All About Astrology

Today I'd like to talk about a subject that's very silly at face value, so silly that anyone with any functional part of a brain laughs it off as childish and ridiculous: Astrology, the notion that the time of year you were born assigns you a zodiac sign, and that sign determines your personality, forecasts your future, and provides decision guidance. The problem is that laughing something off is not really following a very skeptical process, especially when you remember that some important people depend on it. It's appropriate to look at the basis of astrology to see if it has any scientific validity, but it's also appropriate to look at the real-world results to see if there might be some real effect due to a mechanism that's not yet known.

The hardest part about examining the foundations of astrology is trying to determine what they are. It would be nice to be able to at least state that there are 12 signs of the zodiac, but that's only one system. Other astrology systems have 14 or 24 signs. Obviously, they can't all be right.

Even Western astrology with its reliable twelve signs of the zodiac has a serious flaw. Each sign of the zodiac is 30° wide (1/12 of our 360° view of the sky). The precession of the Earth's axis causes our view of the heavens to change over time, and the zodiac used by most astrologers is now wrong by 24°. This means that about 80% of people who have been raised with the understanding that they are of a given birth sign are actually of the preceding sign. My birth date pegs me as a Capricorn, but according to the constellations, I'm actually a Sagittarius. Some astrologers correct for this; others don't. Again, they can't all be right.

Most astrology systems rely on "houses", basically chunks of sky corresponding to each constellation. When a planet moves through a particular house, it's supposed to have a meaning different from when it's in another house. Unfortunately, there are all sorts of varying systems for defining where these houses are (Campaneus, Regiomontanus, and Placidean are the most popular methods in Western astrology), and every astrology system around the world has a completely different interpretation of what the houses mean.

But these only scratch the surface. Most astrological systems are extraordinarily complex, requiring years of study to master, and take many details into account that are far beyond the scope of this book. While it's possible for astrologers to precisely codify exactly how their system is to be used, there are so many different systems, and so many different schools of thought within each, that there are probably as many different ways of doing astrology than there are astrologers. Every single school of thought contradicts another, and every overall system often profoundly contradicts the others. The question "How is astrology done?" has only one right answer: It depends on who your astrologer is.

But differing interpretations don't disprove that there might be some cosmic influence. Whether it's the day, time, or month of your birth, there may indeed be some cosmic force acting upon you that affects your personality. Astrology is prescientific. It was developed millennia before we knew about the actual fundamental forces in nature, thus it makes no claims to having a basis in any real science. That's good, because appealing to any of the real forces in nature would be implausible; each breaks down easily:

- The strong and the weak atomic forces only have a range the size of an atomic nucleus.
- Gravity has an incredibly long range: You are, right now, being affected by the gravity of Neptune. But you're being affected by the gravity from your computer mouse much more. And from your chair even more. And from that person standing in the room even more. Gravity vectors are cumulative. You're

not being pulled this way *and* that way; those vectors add up to a single direction. You cannot decode a gravity vector to determine there was *x* amount of pull from this direction, and *x* amount of pull from that direction. Rearranging the furniture in your room has a gravitational effect on you orders of magnitude higher than the positions of the planets. The moon is the only celestial body that has a significant gravitational effect on us, but most astrological systems give it little or no significance.

- Electromagnetism is the last of the four forces. None of the planets have magnetic fields strong enough to affect us here on Earth. Jupiter's field is the biggest and strongest, but even that is negligible. The sun is the overwhelming electromagnetic source in the solar system, blasting everything else away, but yet again, the sun is not much of a factor in most astrological systems.

In astrology, only the planets, which do not have any detectable effect on us, are assigned all the responsibility for all cosmic influence upon us.

And so, given that there is no detectable effect, you might feel inclined to ask astrologers how they were able to detect its existence themselves, to the point of making it their careers. Generally they'll say they know it's real because it works. Now, I don't want to get into the whole cognitive bias thing here about how people can fool themselves into thinking a metaphysical reading is real; so let's just stick with what we can test and see if astrology really does work.

I wanted to find out if people generally do have the traits that their zodiac signs say they should, and so I conducted an informal survey over Twitter. (To be responsible, I should stress that there was nothing scientific about the way this survey was conducted, and so its results can at best be considered interesting, and not scientific proof of anything.) I went out on the web and found widely available personality descriptions of the various zodiac signs. For example, the words describing a

Sagittarius were generally positive, things that I felt most people would probably identify with. So I took the four phrases (optimistic and freedom-loving, jovial and good-humored, honest and straightforward, intellectual and philosophical) and asked people to assign a 0 to each if they felt it did not describe them at all, a 1 if they felt it somewhat described them, and a 2 if it described them very well. I added each person's points up to get a score from 0 to 8. Since these were generally positive traits, I figured that most people would come up with pretty high scores. The average score turned out to be 6.3, with a clear distribution shoved up to the high end of the graph. The average respondent considers himself a 79% match with the traits of Sagittarius.

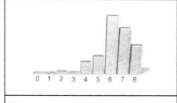

People identifying with Sagittarius traits
0 = Does not describe me at all
8 = Describes me very well
Average: 6.3 (79% match)
Margin of error: 6.2%

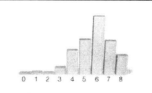

People identifying with Leo traits
0 = Does not describe me at all
8 = Describes me very well
Average: 5.7 (71% match)
Margin of error: 5.2%

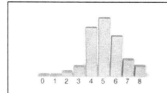

People identifying with Taurus traits
0 = Does not describe me at all
8 = Describes me very well
Average: 5.1 (64% match)
Margin of error: 5.3%

People identifying with Scorpio traits
0 = Does not describe me at all
8 = Describes me very well
Average: 3.7 (46% match)
Margin of error: 4.7%

I also asked the same question using the traits of Leo, Taurus, and Scorpio, adjectives which were (in my estimation) progressively less complimentary, and the average scores did indeed turn out to be 5.7, 5.1, and 3.7. Each of these graphs has a nice, clear bell curve. It's clear that when you ask people, without any context, whether they feel they are better described by words which happen to be positive (like Sagittarius' "optimistic and freedom-loving"), they tend to identify with those terms; but when you ask the same question with less positive words (like Scorpio's "determined and forceful") there is less identification. Armed with knowledge of this fairly obvious axiom, any astrologer should have no problem writing fortunes for just about anyone that will hit the mark 9 times out of 10.

In my survey, I also wanted to see how the results of these same questions might differ between people who are of that zodiac sign, from those who are not. I took the negative qualities of a Libra (indecisive and changeable, gullible and easily influenced, flirtatious and self-indulgent) and asked Libras if they thought it represented them, and asked the same question of non-Libras. If there's anything to astrology, the Libras would have recognized their own weaknesses in those descriptions. But guess what; they didn't. Both groups reported an average of 2.0 out of 6 points, or about a 33% match.

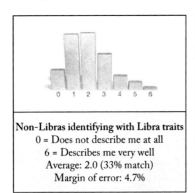

Non-Libras identifying with Libra traits
0 = Does not describe me at all
6 = Describes me very well
Average: 2.0 (33% match)
Margin of error: 4.7%

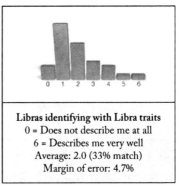

Libras identifying with Libra traits
0 = Does not describe me at all
6 = Describes me very well
Average: 2.0 (33% match)
Margin of error: 4.7%

This result was interesting, so I extended this line of investigation, and asked the same question again, but this time instead of using the zodiac sign's traits, I used randomly chosen readings from the Los Angeles Times horoscope. The first was for Capricorn, and it said:

The universe is sending out some muddled messages. Don't read too much into the signs. If you have to stretch to figure out what something means, it's just because you're not meant to know yet.

Neither Capricorns nor non-Capricorns felt that fortune applied to them much at all; the graphs look virtually identical with a big tall bar in the "Does not apply to me at all" column and only a smattering of results in the other two. Non-Capricorns reported a 12% match, and Capricorns reported a 17% match. While that five percentage point difference may seem significant, it's below the 7.8% margin of error that I calculated for this question.

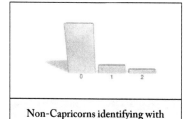
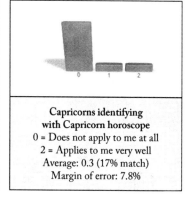

Non-Capricorns identifying with Capricorn horoscope
0 = Does not apply to me at all
2 = Applies to me very well
Average: 0.2 (12% match)
Margin of error: 7.8%

Capricorns identifying with Capricorn horoscope
0 = Does not apply to me at all
2 = Applies to me very well
Average: 0.3 (17% match)
Margin of error: 7.8%

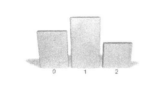

Non-Taurus identifying
with Taurus horoscope
0 = Does not apply to me at all
2 = Applies to me very well
Average: 0.9 (45% match)
Margin of error: 5.9%

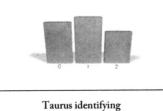

Taurus identifying
with Taurus horoscope
0 = Does not apply to me at all
2 = Applies to me very well
Average: 0.9 (46% match)
Margin of error: 5.9%

I tried it again with a fortune for Taurus that sounded more positive:

> You can do more than instruct people. You can inspire them. You focus on a beautiful potential and describe it with the passion that gets others moving in the same direction.

As you can probably guess, more people felt this applied to them, and this time it made almost no difference at all whether they were Taurus (46% match) or anything else (45% match). Grant a positive sounding fortune, and more people convince themselves it applies to them.

But my little Twitter survey is hardly the first time anyone has tested astrology. Many, many studies have been done; and better the study, the less of an effect has been found. Dutch researcher Rob Nanninga wrote:

> One of the best examples was conducted by the Australian researcher Dr. Geoffrey Dean... Dean selected 60 people with a very high introversion score and 60 people with a very high extraversion score. Next, he supplied 45 astrologers with the birth charts of these 120 subjects. By analyzing the charts the astrologers tried to identify the extroverts from the introverts. The results were very disappointing. It was as if the astrologers had tossed coins to determine their choices. Their average success rate was only 50.2 percent.

He devised his own test in which seven people from diverse backgrounds filled out detailed questionnaires about themselves, and separately provided a list of the dates of important events in their lives. 50 experienced astrologers agreed to match the questionnaires to the date charts, and were offered 5,000 guilders if they could correctly match all seven. As a control, Nanninga also had a group of skeptics try to perform the same matches, to rule out successes based on subtle clues in the data.

The astrologers were asked to indicate how many correct matches they would have expected... Half of them predicted that they had matched all subjects with the correct charts. Only six astrologers expected less than four hits. In fact, the most successful astrologer achieved only three correct matches, whereas half of the participants (22) did not score a single hit. The average number of hits was 0.75...

Moreover, there was no evidence that the most experienced astrologers did any better than beginners. It is interesting to compare the entries of the participants with each other. Because they all had received the same information, one would expect many similar responses. Actually, the lack of agreement was striking. Each of the seven charts could be paired with seven questionnaires. Of these 49 possible combinations, none was selected more than twelve times. It was as if each astrologer had used a random generator to determine the correct matches.

Of the control skeptics, the most successful also scored three hits, the same as the best astrologer.

In addition to his 1987 study referenced by Nanninga, Dr. Geoffrey Dean also performed a meta analysis of nearly 300 empirical studies of astrology. He found no real effect, and attributed the perceived effect to perceptual and cognitive biases that he called hidden persuaders. In his conclusion he wrote:

To critics, astrology's failure to deliver is unremarkable because its alleged efficacy is explained by the same hidden persuaders that underlie proven invalid approaches such as phrenology and bloodletting... Each hidden persuader creates the illusion that astrology works, all are used routinely in consulting rooms, all lead to client satisfaction — and none require that astrology be true... If you are looking for something where nothing is true and everything is permitted, then astrology seems to be an excellent choice.

People who are big believers in their horoscopes are probably going to continue to remain so, no matter how much evidence they're shown that any perceived effect is imaginary. But for those who are on the fence, this information is crucial, given that some leaders in government and business employ astrology in their decision-making. If someone with authority over your life is making important decisions based on magical beliefs, you should trust the science, not the authority.

References & Further Reading

Culver, R., Ianna, P. *The Gemini Syndrome: A Scientific Evaluation of Astrology.* Buffalo: Prometheus Books, 1984.

Dean, G. "Meta-analyses of nearly 300 empirical studies: Putting astrology and astrologers to the test." *astrology-and-science.com.* Astrology & Science, 28 Sep. 2008. Web. 28 Sep. 2009. <http://www.rudolfhsmit.nl/d-meta2.htm>

Grim, Patrick, editor. *Philosophy of Science and the Occult, Second Edition.* Albany: State University of New York Press, 1990. 15-84, esp. 51-81.

Jenni. "Astrology is Bunk." *Debate Unlimited by Martin Willett.* Debate Unlimited, 14 Jan. 2008. Web. 28 Sep. 2009. <http://mwillett.org/mind/astrology-is-bunk.htm>

Nanninga, Rob. "The Astrotest: A Tough Match for Astrologers." *Correlation, Journal of Research into Astrology.* 10 Feb. 1997, Volume 15, Number 2: 14-20.

49. More Medical Myths

Once before we looked at medical myths perpetuated by movies and pop culture, but the sheer volume of misinformation can hardly be contained within a single chapter. So today we're going to pick it up again, and look at some more stories about the human body that you've always heard and probably believe.

We'll begin with the age-old advice that you should drink at least eight glasses of water a day, or about two liters. If you're backpacking or bicycle racing, that's not really all that much. But for most people, who, like me, sit around watching TV and scratching their belly, this would merely lead to superfluous trips to the bathroom. The problem is that the advice is not only unsupported, it's also misrepresented. The original recommendation seems to come from the Food and Nutrition Board of the National Research Council, which, way back in 1945, said that you should do this. But what seems to have been forgotten is that the report added, "Most of this quantity is contained in prepared foods." Omit that, and it appears that you're supposed to stand at the sink and fill your glass eight times a day; which, it turns out, nobody ever recommended in the first place. Whatever you drink normally, in the form of coffee, juice, soft drinks, whatever, probably satisfies most people's water requirements. The evidence for this is that we don't see suited businesspeople laying around on the sidewalks of New York City dying of thirst, stretching out their hands in appeal to passersby.

Just don't take this advice to the extreme. My dad and some friends once came home from college in the UK by buying a boat and sailing across the Atlantic, and they needed to stock up on drinking water. But they reasoned that any liquid was just as good, so they bought a couple kegs of wine instead of water. The wine quickly spoiled in the tropical doldrums and had to be discarded, leaving them stranded in the middle of the Atlantic with nothing to drink, and by the time they reached the New World they were practically dead. Moderation is always better than complete reliance on anything.

Dying from thirst is just one thing that can be frightening. They say Marie Antoinette's and Sir Thomas More's hair turned white from terror the nights before they were executed. It also happened in Shakespeare, and it even happened to Jobeth Williams in *Poltergeist,* giving her some Cruella de Ville streaks of white overnight. Can a terrible shock turn your hair white? More to the point, can *anything* make existing hair strands change color? Discounting artificial coloring or sun bleaching, the answer is no. Hair is dead tissue, there is no metabolism or mechanism that could alter its pigmentation and change its color, no matter how big of a shock you receive. However there is a condition called alopecia areata, in which all your pigmented hair falls out, sometimes quite rapidly, leaving only any unpigmented hair you might have. On a person with salt and pepper hair, this could indeed have the apparent effect of turning your hair white overnight. But although the causes of alopecia areata are unknown (though it's suspected to be an immune disorder), attempts to link it to stressful events have been post-hoc rationalizations, and there is no good evidence that it can be caused by fright or stress.

But what about new hair growth? Can some frightening event change something in your body that changes the color of your new hair? People often talk about stressful relationships or projects giving them gray hairs; does this actually happen? Not really, no. The color of your hair, and the age at which it goes gray, is determined by your genes. However, the process is not completely uninterruptible. Certain chemotherapies and certain diseases can temporarily change the color of your new hair growth, but your color will return to normal after the episode. Ionizing radiation has been shown to bring on premature graying in mice due to genetic damage. But so far, no good evidence supports the idea that a sudden fright, or even years of stressful living, can prematurely cause your hair to start growing gray or white, or cause hair loss.

And what man wants to be bald or gray when he's trying to pick up women? We've all heard that men think about sex every eight seconds. Or seven seconds, or nine seconds, or whatever the number. They think about it a lot. And if you're a man, you've probably heard this, felt ashamed and less of a man, and concluded you need to get your game on and think about sex a lot more often just to keep up with your peers. This is the kind of thing that sounds like it must have come out of some study done in the 1970's. The fact is nobody's really sure where or how this claim got started. Researching the question usually yields the same source: A study published in 1994, *The Social Organization of Sexuality: Sexual Practices in the United States.* They conducted a large survey, and although the results depended on self-reported data by men, they found that 54% of men think about sex at least once a day, 43% at least once a month, and 4% less than once a month. Hundreds of times a day does not seem to be supported by any real data. So relax, your hormones are probably OK.

So since you're not obligated to think about sex all day long, you'll need to find some other way to get knocked out. In movies it's really easy to knock someone out. If you're Bones McCoy, a couple quick karate chops to the shoulder and abdomen will do it; for anyone else, a solid punch or a sharp crack to the back of the head with a pistol will instantly drop your adversary to dreamland. 20 minutes later, they come to, perhaps feeling a bit bruised, but otherwise uninjured. Do people really have such an easily accessed and inconsequential on/off switch?

The actual injury needed to knock someone out is called a mild traumatic brain injury, or MTBI, more commonly called a concussion. Concussions are not caused simply by any blow to the head, though such a trauma can certainly cause other injuries. Concussions are produced by rapid brain acceleration, deceleration or rotation, anything that rapidly gives your brain a good squish. Grade I and Grade II concussions produce confusion and/or amnesia for up to 24 hours, but it takes a Grade III concussion to actually lose consciousness, and this rarely lasts more than five minutes. Symptoms sure to follow include headache, dizziness, confusion, nausea, slurred speech, lack of reasoning ability, amnesia and disorientation. In most patients, the symptoms resolve themselves within a matter of weeks. Rarely, surgery may be needed to relieve pressure or fix an intracranial hemorrhage. So if you're looking for that quick on/off switch for your movie plot, you need to invent your own imaginary clinical reactions.

But once you're unconscious, won't you get fat if you've recently eaten? Sumo wrestlers are famously said to maintain their large bulk by napping after their meals. The idea is that food you eat before sleeping doesn't get as thoroughly metabolized as food you digest while being active, and so if you want to get fat, shift your meals toward bedtime. Lots of people try to avoid eating after dinnertime to stay skinny, but does this really do any good? Sadly for those looking for easy answers to the weight gain problem, this particular solution has been studied a lot, and found to be just another myth. There's certainly a correlation between people who eat late at night and obesity, but this is simply because people who eat at night are more likely to be those who overeat throughout the day. Similarly, those who skip breakfast are more likely to be those who power away too big of a lunch. Turns out it doesn't matter when you eat your calories, it just matters how many calories you eat in total. People who eat small meals spaced throughout the day are less likely to overeat at any given meal. So relax and don't shy away from that late-night snack, just be aware during your other meals that more calories are coming later.

But while you're snacking in bed, is it safe to pick up a book? Mom always used to warn me "Don't read in the dark, you'll ruin your eyes." And so, like most people, I've tried to avoid this. But then an ophthalmologist friend assured me

"You can't hurt your eyes by using them." Which do you trust, medical science or Mom wisdom? Can reading in dim light damage your visual acuity? There are two lines of evidence that support this: First, that using your eyes in difficult conditions can cause discomfort in the form of eyestrain. That's a fact. Second, that smart people tend to wear glasses, presumably meaning that people who read a lot to achieve academic stature damaged their eyes in the process. That one? Not so much of a fact. Again, there may be an actual correlation: People who need to read a lot in their profession may be more likely to have gotten glasses to facilitate their reading, but there's no good evidence that one causes the other. Eyestrain is not cumulative, and once you stop reading, the strain goes away and your eyes return to normal. It should be noted, however, that when you research this, you will find articles that do support the claim. But they are very much in the minority. It doesn't make them wrong, it just means that the experimental data has led the majority consensus of researchers to the opposite conclusion.

I say take your snack to bed with you, and read in the dark. If thinking about sex doesn't put you to sleep, a few sudden well-placed brain decelerations from smacking your head on the wall should do the trick. If it doesn't, then you should have known better than to listen to me or to anyone else. Always think for yourself, and never blindly accept what pop culture tells you.

References & Further Reading

Hofmekler, Ori. "Diet Fallacy #3. Eating late will make you fat." *DragonDoor.com For supreme fitness and well being.* DragonDoor.com, 12 May 2005. Web. 9 Jan. 2010.
<http://www.dragondoor.com/articler/mode3/318/>

Jelinek, J. E. "Sudden whitening of the hair." *Bulletin of the New York Academy of Medicine.* 1 Sep. 1972, Volume 48 Number 8: 1003-1013.

Kim, Ben. "Why Drinking Too Much Water Is Dangerous." *Dr. Ben Kim's Blog.* Dr. Ben Kim, 31 Mar. 2009. Web. 9 Jan. 2010.
<http://drbenkim.com/drink-too-much-water-dangerous.html>

Laumann, Edward O., Gagnon, John H., Michael, Robert T., Michaels, Stuart. *The social organization of sexuality: sexual practices in the United States.* Chicago: The University of Chicago Press, 1994.

Levy, Janey. *Alopecia Areata.* New York: The Rosen Publishing Group, 2007.

Pearce, J.M.S. "Observations on Concussion." *European Neurology.* 1 Feb. 2008, Volume 59 Number 3-4: 113-119.

Ropper, Allan H., Gorson, Kenneth C. "Concussion." *The New England Journal of Medicine.* 11 Jan. 2007, Volume 356 Number 2: 166-172.

50. Shadow People

They usually come at night. Maybe you're reading or watching TV or just laying in bed. He's most often a man, and may be wearing a hat or a hood. A lot of times you'll only catch a glimpse of him out of the corner of your eye, as he flits across the wall or disappears through a doorway. Sometimes he's just a shadow, a flat projection sliding across the wall or ceiling; but other times, especially in the dark when you least expect it, shadow people appear as a full-bodied black apparition, jet black like a void in the darkness itself, featureless but for their piercing empty eyes.

The foggy Santa Lucia Mountains run along the central coast of California, and for hundreds of years, the Chumash Indians and later residents have told of the Dark Watchers, shadowy hatted, caped figures who appear on ridges at twilight, only to fade away before your very eyes. A visit to the Internet reveals hundreds and hundreds of stories from people who saw shadow people in their homes, on web sites such as shadowpeople.org, from-the-shadows.blogspot.com, and ghostweb.com:

> *I opened my eyes and looked towards the middle of the room. I saw a large shadow in the shape of a person. It had no facial features that I could see and it wasn't moving. It was just standing there looking at me... I blinked and then it was gone.*
>
> *I felt like someone was watching me so I turned to look toward the hallway and there it was in the doorway... It was a black figure. I could only see from the torso up. I felt it was a male and could feel that it was looking at me... I started to walk towards it and it disappeared back into the room.*
>
> *There, at the foot of my bed, was a tall dark figure like a shadow. It appeared to be almost 7 feet tall with broad*

> shoulders and was wearing what seemed to be an old fashioned top hat and some sort of cape... I watched as it glided past me and out the door of my room.

It goes without saying that skeptics have long-standing explanations that, from the comfort of your armchair, adequately rationalize all the stories of shadow people. These explanations run the gamut, all the way from mistaken identification of a real shadow from an actual person or object, to various causes of optical illusions or hallucinations like drugs or hypnogogic sleeping states, even simply lying and making up the story. I think that probably everyone would agree that these have all happened, and therefore they do explain some people's experiences. But here's a fact: Try to offer any of those explanations to someone telling you about a specific sighting, and it will likely be immediately shot down. "I was not asleep." "I know the difference between a regular shadow and what I saw." "What about my friend who saw it with me?"

The truth is that it's probably not possible to explain most sightings. If it *was* some mysterious supernatural noncorporeal being who flitted through the room, no evidence would remain, and thus there's nothing to test or study. It's so trivial to fake photos or video of something as vague as a shadow person that when these exist, they're interesting but practically worthless as far as empiricism goes. Only in the rare case where an actual physical cause can be found, and you're able to consistently reproduce the effect at the right location and the right time of day and in the right lighting conditions, are you able to provide a convincing explanation. Most of the rest of the time, all you have is conjecture and hypothesis, and the eyewitness is likely to reject these.

When I was a kid we once lived in a house where if you walked up the stairs and one of the upstairs bedroom doors was open a crack, you might see a flash of movement inside the room from the corner of your eye. I saw it a number of times, and other people in my family did too. I thought it looked like someone threw a colored sweatshirt across the room. But: I *never* saw it whenever I walked carefully up the stairs and kept my eyes on that crack; it only happened if you weren't looking right at it and weren't thinking about it. The more you learn about how the brain fills in data in your peripheral vision and blind spots, the less unexpected and strange this particular experience becomes. I have no useful evidence that anything unusual happened, and I have good information that can adequately explain what was perceived. I personally am not impressed enough to deem it worthy of further investigation, but others might be, and that's a supportable perspective. But unless and until some substantial discovery is made, the determination that it must have been a shadow person or ghost is ridiculous. Nothing supports that conclusion. And yet my story is at least as reliable as 99% of the shadow people stories out there. I was not on drugs, I know the difference between a shadow and what I saw, and other people saw it too.

Enthusiasts of the paranormal offer their own set of additional hypotheses about shadow people. One proposes that shadow people are the embodiments of actual people who are elsewhere but engaged in astral projection. This is not an acceptable hypothesis. Like shadow people themselves, astral projection is an untestable, undetectable, unprovable conjecture. Explaining one unknown with another unknown doesn't explain anything, and the match itself cannot be made, since neither phenomenon has any known properties that you could look at and say, "What we know of shadow people is consistent with what we know of astral projection." We know nothing about either, so there's no logical basis for any connection.

The same can be said of another paranormal explanation for shadow people, that they are "interdimensional beings". Let's make an outrageous leap of logic and allow for the possibility that interdimensional beings exist. What characteristics would they have? How would we detect their presence? What level of interaction would they have? How would they affect visible light? Since these questions don't have answers, you can't correlate interdimensional beings to the known properties of shadow people. Neither one has any.

But there are phenomena to which we can correlate these stories. We know the details in the eyewitness accounts, and we know the psychological manifestations of conditions like hypnogogia and sleep paralysis. A hypnogogic hallucination is a vivid, lucid hallucination you experience while you're still falling asleep. You're susceptible again eight hours later when you're waking up, only now it's called hypnopompia. But this seems such a cynical, closed-minded reaction. When you suggest hypnogogia as a possible explanation to a person who has witnessed shadow people, many times their reaction will be understandably negative, if not outright hostile. "You're saying I'm crazy" or "You're saying I imagined it" are common replies. Hypnogogia is neither a mental illness nor imagination, and to dismiss it as either is to underestimate the incredible power of your own healthy brain. Too many people don't give their brains enough credit.

I had a dramatic demonstration of the power of hypnopompia — the waking up version — when I was about 10 years old. Early one morning, the characters from *Sesame Street* put on a show for me in the tree outside my bedroom window. It had music, theme songs, lighting cues and costume changes: A full elaborate production, and it lasted a good hour. To this day, I have clear memories of some of the acts. I even went and woke my parents to get them to watch, but by then the show had gone away. I knew for a fact that I hadn't been asleep. I'd been sitting up in bed and writing down some of the songs they sang. Those writings were real, on real paper, and even made sense when viewed in the light of day. It had been a completely lucid, physical experience for me. *But it only existed inside my own brain in a hypnopompic state.* My brain had composed music, performed the music, written lyrics, and sang them in silly voices for some director who must also have come from within me. The skits were good. The actors were rough-sewn muppets, independently moving and climbing about, even swinging through the swashbuckling number, on tree branches representing the lines of a great pirate ship. Yet through it all, I'd been conscious and upright enough to actively transcribe the lyrics. *That's* the power of a brain.

But many believers reject the idea that their brain has such capabilities, and instead conclude that any such perceptions can only be explained as visitations from supernatural entities. One such believer, Heidi Hollis, has gone on Coast to Coast AM radio a number of times with suggestions to defend yourself from shadow people:

- Learn to let go of your fear.
- Stand your ground and deny them access to your person.
- Focus on positive thoughts.
- Use the name of Jesus to repel them.
- Keep a light on or envision light surrounding you.
- Bless your room with bottled spring water.

Interestingly enough, such actions may actually work (although it's not the techniques themselves that are responsible — plucking a chicken or beating a drum could work just as well, if you think it will). Sleep disorders in the form of disruptive episodes such as these are called parasomnias, and the primary treatments for parasomnias are relaxation techniques, counseling, proper exercise, and the basic lifestyle changes that contribute to better sleeping habits. True believers who reject any notion suggesting their experience was anything but a genuine visit from a supernatural being, but who apply any such remedies as Hollis suggests, do indeed have a good chance of finding relief, when the process of applying the remedy brings them some peace of mind. Even though these remedies are rarely going to be as effective as professionally guided treatment, the fact that they can sometimes work only reinforces the true believers' notion that the shadow person was in fact an interdimensional demon, and that sprinkling holy water around the room did in fact scare it away.

These experiences are weird, and can be scary. But they're also fascinating, once-in-a-lifetime opportunities to experience the true power of your brain. To conclude that it's a supernatural being is to rob yourself of the real wonder of what's probably happening. Faith in the supernatural offers you nothing better than an implausible and ignorant supposition that stifles further understanding, while the willingness to accept science gives you a whole universe without limits.

References & Further Reading

Bell, Carl C. "States of Consciousness." *Journal of the National Medical Association.* 1 Apr. 1980, Volume 72, Number 4: 331–334.

Bishop, G., Oesterle, J., Marinacci, M., Moran, M., Sceurman, M. *Weird California.* New York: Sterling Publishing Company, Inc., 2006. 56.

Guthrie, S. *Faces in the Clouds.* New York: Oxford University Press, 1993. 91-121.

Hines, T. *Pseudoscience and the Paranormal.* Amherst, NY: Prometheus Books, 2003. 91-93.

Schlauch R. "Hypnopompic Hallucinations and Treatment with Imipramine." *American Journal of Psychiatry.* 1 Jan. 1979, Volume 136: 219-220

INDEX

2001 A Space Odyssey.... 153, 154
9/11 attacks....59, 61, 88, 115, 159, 161
Abraham, Book of ...271, 272, 273, 274, 275, 276
acupuncture 63, 116, 117, 186
Agincourt, Battle of......... 121
Ahmadinejad, Mahmoud 228, 230
Aladdin........................... 127
Ali Baba........................... 127
alien abduction..............66, 70
alkaline water.................. 143
alopecia areata.................. 304
Amityville Horror, The 31
AMORC ..244, 247, 248, 249
Amos, Wally (Famous Amos) 252
Angel of Mons..120, 122, 124
Animal Liberation Front (ALF)............................ 60
Apollo program 59
Architeuthis...................... 191
Arecibo, PR 52
Arnold, Chris, Airman USAF 109
aspartame...40, 41, 42, 43, 44, 45, 46
astral projection 311
astrology.....86, 294, 295, 296, 298, 300, 301, 302
Atiya, Aziz........................ 272
Atkinson, William Walker 247, 248, 249

Atlantic Richfield Company (ARCO) 54, 55
aura 35, 63, 149
aurora 53
Austin, Steven . 165, 167, 168, 170
autism................................ 62
autoimmune diseases.. 43, 142
Bacon, Sir Francis 78, 245
Bandler, Richard 203, 204
Barakat, Ali 150, 151
Barker, Gray.................... 222
Barnum, P.T. 136
Battle of Los Angeles (1942) 24, 25, 26, 27, 28
Bauer, Eddie 254
Bearden, Thomas ... 49, 50, 51
Begbie, Harold................ 124
Begich, Nick 55
Bell Witch. 29, 30, 31, 32, 33, 34
Bell X-1 Glamorous Glennis 195, 196, 197, 200, 201
Bell, Allen 31, 32
Bell, Art 35, 36
Bell, Elizabeth.............. 29, 30
Bell, John ... 29, 30, 31, 32, 33
Bell, Richard 31, 32
Bentwaters (Royal Air Force base)...................... 105, 111
Beowulf........................... 126
Bernadette (St. Bernadette Soubirous)................ 71, 72
Better Business Bureau 61
binaural beats .. 171, 172, 173, 174, 175, 176

Bioelectromagnetics (journal) 15
Blair Witch Project, The 30
Blaylock, Russell.................42
bog people 74
Bohemian Club 87, 88, 89, 90, 91, 92, 93
Boiardi, Ettore (Chef Boyardee) 251, 256
Bondarenko, Valentin..........8
Boone, Daniel...................127
Borley Rectory 31
Bosnia...............146, 151, 152
Bowmen, The...121, 122, 123
Brainiac (TV show)............15
Bravewell Collaborative......62
Brigham Young University 273, 274
British Expeditionary Force 120
brown dwarfs 157
Browne, Sylvia....................58
Buddhism 74
Bunyan, Paul 128
Buran, Fred, Lt. USAF....108
Burroughs, John, Airman USAF...................108, 109
Cabansag, Edward, Airman USAF...................108, 109
Cable News Network (CNN) 12, 13, 14, 15, 17, 61
Callender, Marie & Don ..253
cancer...............13, 16, 17, 64
Cardiff Giant 136
Cardo Systems.....................15
Cassini (spacecraft)..154, 155, 156, 158
Castus, Lucius Artorius 129

Catherine (St. Catherine of Bologna) 72
Catholic church......71, 74, 75
cellular phones....................14
Chapman, John (Johnny Appleseed)............127, 131
Charles, Prince of Wales...62, 229, 230
Charteris, John, Brig. Gen. 122
China 8, 20, 56, 278, 279, 280, 281, 282
Clarke, Arthur C.......153, 157
Clarke, David............122, 123
climbing perch 21
Coast to Coast AM 35, 36, 37, 38, 155, 224, 313
Cohen, Ben.......................251
collagen 191, 192, 193
Columbus, Christopher 94
concussion 305
Connolly, Robert 133
Conservapedia. 209, 210, 211, 212, 213, 214
conspiracy theories 28, 42, 43, 44, 50, 51, 53, 54, 55, 56, 59, 61, 63, 87, 88, 89, 91, 92, 153, 154, 155, 156, 159, 160, 161, 164, 215, 219
Cooper, Tim 26, 184
Coral Castle 177, 178, 179, 180, 181, 182
Coronado, Rod 61
Cosmos (satellite program) 107
Cowdery, Oliver 272, 274, 275
Crocker, Betty..................251

Crusoe, Robinson............ 129
Cultural Revolution (China)
.... 280
dacite........165, 167, 168, 170
Dalai Lama......277, 278, 280, 281, 282
Dark Watchers................ 309
Darwin, Charles............... 209
Davis, Debra.................12, 13
daylight saving time.288, 289, 290, 291, 292
Dean, Geoffrey.........300, 301
Dean, Jimmy..................... 252
Defense Advanced Research Projects Agency (DARPA)
............................53, 54, 57
DeFoe, Daniel................ 129
dehydration..........75, 303, 304
Deneb (star)....................... 84
detoxification................42, 63
Diego Garcia island......... 162
dielectric effect................... 56
disaccharides..................... 216
Discovery Channel.......... 105
Disney, Walt.................... 210
Dittmar, Heini..........197, 201
Domhoff, William........91, 93
Doyle, Sir Arthur Conan. 131
Drake, Sir Francis.............. 78
Dula, Tom (Tom Dooley) 128
Earth Liberation Front (ELF)
............................53, 57, 60
Eastlund, Bernard...54, 55, 57
Edell, Dean....................... 41
EISCAT........................... 52
electroencephalogram173, 174
electrolysis..........140, 145, 285

embalming............. 73, 74, 75
Encyclopedia Britannica....19
entrainment............. 172, 173
Erickson, Milton...... 204, 205
Explorer I............................ 6
Faktoriwicz, Maximilian (Max Factor)................ 254
farming...... 79, 108, 112, 232, 233, 235, 253, 258, 259, 260, 263, 264, 289, 290
Fata Morgana mirage...... 102, 103, 104
Federal Bureau of Investigation............. 26, 28
Federal Emergency Management Agency (FEMA)....... 159, 160, 161
Ferguson, Thomas Stuart. 273
fertilizer.................... 259, 260
Fields, Debbi (Mrs. Fields)
.... 252
fish........... 18, 20, 21, 22, 259
Fish, Marjorie.................... 69
Food and Drug Administration.......... 41, 46
formaldehyde.......... 40, 44, 73
Franklin, Benjamin. 163, 288, 293
Freeman, Gordon............... 37
Friedman, Stanton...... 25, 28, 117, 118
frogs..................................20, 21
fructose.... 215, 216, 217, 218, 219, 220
Fulford, Benjamin.............. 56
fungicide......................... 261
Gagarin, Yuri.............. 8, 9, 11
Galileo (spacecraft)........... 154

Garden Plot 159, 162
Geocron Laboratories 167
Gliese Star Catalog 69
Globsters 190, 192, 194
glucose 216, 217, 218, 220
Goodspeed's History of
   Tennessee 32, 34
gray aliens 65, 67, 132
Great Leap Forward (China)
   280
Greenfield, Jerry 251
Greenpeace 96
Grendel 126
Grinder, John 204
guaiacol 73
Guantanamo Bay 162
Half-Life 37
Halliburton 161, 163
hallucination 312
Halt, Charles, Lt. Col. USAF
   .... 106, 110, 111, 112
Hatfields (clan) 128
Hawass, Zahi 150
Heap, Michael 207
Hedlock, Reuben 272, 274
Henderson, Paul 65
Henry, Bill 27
Henry, John 127, 131
High Frequency Active
   Auroral Research Program
   (HAARP) ... 52, 53, 54, 55,
   56, 57
high fructose corn syrup .. 215, 217
Hilfiger, Tommy 255
Hill, Barney 65, 66, 67, 68, 70
Hill, Betty 66, 67

Hill, Ernestine 101
HIPAS 52
History Channel 105
Hitchcock, Alfred 239
Hitchens, Christopher 281
Hitler, Adolf 42, 177, 209
Hoagland, Richard 155
Holmes, Sherlock 131
Holocaust 60
homeopathy 63, 186
Homer 126
Hood, Robin 129
Hope, Dennis 83
Hopewell tradition 138
Hurricane Katrina 49, 161
Husted, Marjorie 251
Huygens, Christiaan 172
hydrocephaly 132
hypnogogia 310, 312
hypnopompia 312
I-17 (Japanese submarine) .. 24
Iliad, The 126
Ilyushin, Vladimir 8, 11
incorrupt corpses ... 71, 72, 73, 75, 245
Ingram, Martin Van Buren 31, 32, 33, 34
Insurrection Act 160
International Astronomical
   Union 84, 85, 86
International Star Registry 82, 83
ionized water 142
ionosphere 53, 54, 55, 57
Itigelov, Hambo Lama
   Dashi-Dorzho 74, 75

Jackson, Andrew 29, 30, 33
Jakhu, Ram 84
James, T. Horton 99
Japan .. 24, 28, 47, 56, 74, 162, 163, 174, 190, 192
Jones, Alex 88
Jones, Davy 130
Jones, John (Casey) 128
Judica-Cordiglia, Achille and Giovanni 5, 6, 7, 8, 9, 10
Jupiter (planet). 140, 153, 154, 157, 158, 296
Kangen water filters. 140, 141, 142, 143, 144
Keel, John 222
Kennedy, John Fitzgerald. 59, 131
Kidd, William (Captain Kidd) 77, 78
King, Larry 14, 61
Klass, Philip 26, 28, 70
Knox, Frank 25
Koelbjerg Woman 73
Laika 6
Lamond, Henry 100, 103
Lebolo, Antonio 271, 273
Leedskalnin, Edward 177, 179, 181
Lee-Enfield (rifle) 121
Lippisch, Alexander .. 196, 198
locally grown food 232, 233, 234, 235
Loma Prieta earthquake (1989) 20
Los Angeles Times 25, 214, 261, 299
Louis XIV (king) 130

Lower, Stephen 141
Lucifer Project 153, 154, 155, 157, 158
mach jump 198, 200
mach tuck 198, 199
Machabeli, Giorgi (Prince Matchabelli) 254
Machen, Arthur 121, 122, 124
Maher, Bill 61, 227, 229
Majestic 12 .... 25, 26, 28
Manhattan Project 91
Manzanar 162, 163
Markle, Nancy 43
Marshall, George . 25, 26, 194
Marshall/Roosevelt Memo (1942) 26
MarshCreek, LLC 54
Mayer, Oscar 253
McCarthy, Jenny .. 62, 64, 229
McCoy (clan) 128
McGillivray, James 128
McGinnis, Daniel 77, 80
McGuffin 239, 242
McKinney, Cynthia 161
McVeigh, Timothy .. 229, 230
mechanoluminescence 102
Messerschmitt Me-163 Komet 196, 197
Messerschmitt Me-262 Schwalbe 198, 199
microwave ovens 14
Min Min light ... 99, 100, 101, 102, 103, 104
Ministry of Defense (UK) 106
Moloch 87, 90
Mons, Belgium . 82, 120, 121, 122, 123, 124

Moore, Charles..................97
Moore, Michael.............215
Moravec, Mark.........100, 104
Morgan, Henry..................78
Mormon, Book of.....271, 273
Mormonism.....................276
Mount St. Helens....165, 168, 169, 170
mucoid plaque 143
multilevel marketing 140, 141, 144, 240, 242, 254
mummies..........71, 72, 74, 75
Mutke, Hans Guido 198, 199, 200, 201
Muzak 174
Nanninga, Rob .300, 301, 302
National Aeronautic and Space Administration (NASA) 154, 155, 156, 158
National Cancer Institute..13, 16
National Enquirer.............117
National Oceanic and Atmospheric Administration...96, 97, 98
National Science Foundation 50
Nature (journal)........102, 137
Nepal................278, 279, 281
neuro-linguistic programming 203, 204, 205, 206, 207, 208
neuston plastic 97
new age.......................55, 63
New World Order.............160
Nibley, Hugh....................274
Nickell, Joe 76, 80, 81, 224, 225

nocebo effect...................186
Noory, George 37
Norris, Chuck.....................59
North American Air Defense Command (NORAD)...10
North American XP-86 Sabre 200, 201
Norton, Mary....................120
Novella, Steven ... 1, 132, 139, 270, 287
Oak Island, Nova Scotia ...77, 78, 79, 80, 81
Obama, Barack.................161
Oberg, James.................8, 11
Octopus giganteus... 189, 191, 193, 194
Odor of Sanctity.................73
Odysseus 126
Odyssey, The 126
Office of Air Force Hisotry 25, 27, 28
Office of Naval Research (US Navy) 54
Oldfield, Harry 149
Olympus Mons 82
One Thousand and One Nights 126
Orfordness lighthouse..... 108, 109, 111, 112
organic food 216, 232, 257, 258, 259, 260, 261, 262, 263, 264
Osborne, Gerald (Red Elk) 36
Osmanagić, Semir... 146, 147, 148, 149, 150, 151
Ouellette, Jennifer.... 242, 243
oxidation 143
Oz, Mehmet (Dr. Oz)......217

Pacific garbage patch ...94, 96, 97, 98
Palin, Sarah........226, 227, 229
papyrus.....244, 271, 272, 273, 274, 275, 276
Pearl Harbor...................... 24
Pearl of Great Price 272
Pease Air Force Base 65, 68
peat bogs............................ 73
Pendragon, Arthur (king) 102, 129
Penniston, James, SSgt. USAF..................108, 109
People for the Ethical Treatment of Animals (PETA)....................60, 61
Perls, Fritz........................ 203
Pettigrew, Jack...........102, 104
Pfizer Inc. 42, 116
pH 140, 143, 144
Phelps, William........272, 275
phenylalanine..........42, 43, 45
phenylketonuria................. 45
Pied Piper........................ 130
piezoelectric effect 101, 103
Pio of Pietrelcina (Padre Pio) 73
pirates 77, 78, 79, 313
placebo effect ...183, 184, 185, 186, 187, 188
plutonium 155, 156
Pocahontas................128, 131
Pollan, Michael................ 235
Posse Comitatus Act 160
potassium-argon dating .. 166, 167, 168, 169, 211

Project Blue Book (US Air Force)...... 59, 65, 68, 69, 70
PubMed 219
Pye, Lloyd 132, 133
Pyle, Ernie......................... 27
pyramids.... 63, 141, 146, 147, 149, 151
radiation, extremely low frequency (ELF) 53, 57
radiation, ionizing 304
radiation, radio frequency (RF) 14, 15, 16
radioisotope thermoelectric generator 154, 155, 156, 157
radiometric dating 79, 165, 166, 168, 169, 211
Regulus (star) 84
Religulous (movie) 61
Rendlesham Forest UFO 105, 106, 111, 113
Rennie, John 60, 64
Rex-82 Proud Saber 160
Rex-84 Night Train 160
Ridpath, Ian 110, 113
Rogan, Joe........................ 59
Rolfe, John 128
Roosevelt, Franklin D. . 25, 26
Rosenkreuz, Christian..... 244, 245, 248
Rosicrucianism 244, 245, 246, 247, 248, 249
Roswell Daily Record....... 117
Rumsfeld, Donald .. 41, 42, 44
Salatin, Joel 235, 236
Sargasso Sea 94, 95, 98
sargassum (seaweed)........... 95

Satir, Virginia 204
Saturday Evening Post, The
 32
Saturn (planet). 154, 155, 157, 158
scalar field theory 47, 48, 49
scalar weapons 49, 51
Schlafly, Andrew 212
Science (journal) 95
Sci-Fi Channel 105, 108
Searle, G. D. & Company. 42, 44
Seattle Times 36, 39
shadow people . 309, 310, 311, 312, 313
Shandera, James 25
shock stall . 196, 197, 198, 199
Sichuan earthquake (2008) 20, 56
Silvan (St. Silvan) 72
Sinbad 126
Sirius (star) 84
skeletons 134, 135, 137
Skeptical Inquirer (magazine) 28, 67, 81, 225
SkepticBlog.org 232, 234, 270
skepticism 51, 80, 115, 163, 238, 242, 277, 284, 287, 288
skull binding 133
sleep paralysis 312
Smailbegovic, Amer 148
Smith, John 128
Smith, Joseph .. 271, 272, 273, 274, 275, 276
snakehead (fish) 21
Snopes.com 41

sokushinbutsu 74
solar radiation 52, 53
Soviet Union . 5, 6, 7, 8, 9, 10, 11, 47, 157
sperm whale 193
Sputnik (satellite program).. 6, 9
St. Augustine Monster 189, 191, 192, 193
Stein, Ben 60, 64, 229
Stone, Oliver 59
stratosphere 53, 195
stress. 183, 187, 258, 296, 304
sucralose 45
sucrose 216, 219, 220
sugar 33, 42, 102, 215, 216, 217, 218, 219
supersonic flight 196, 197, 198, 199, 200, 201
Sura 52
Talmud 133
Teach, Edward (Blackbeard)
 78
Tell, William 130
Teller, Edward 91
Templar knights 244, 245
Tereshkova, Valentina 9
Tesla, Nikola 47, 48, 51
The Outer Limits 67, 68
thermal inertia 148
Thurkettle, Vince 110
Tibet 277, 278, 279, 280, 281, 282
Tolkien, J.R.R. 120
tornados 19, 20
Torre Bert 6, 7, 8, 10
troposphere 53

Trudeau, Kevin............... 242
Truman, Harry................. 25
Tsar Bomba..................... 157
Tunguska event (1908).47, 51
TWA Flight 800 49
Twain, Mark.....131, 135, 136
Tyrannosaurus rex........... 212
Ulysses............................. 126
unidentified flying objects. 24, 25, 26, 27, 28, 61, 65, 67, 68, 70, 105, 106, 107, 113, 117, 222, 225
United States Air Force.... 25, 27, 28, 53, 65, 68, 69, 105, 106, 109
United States Navy.24, 25, 53
urushi tree......................... 74
V-2 (rocket program)....... 196
van der Worp, Jacco......... 155
Vega (star) 84
Verne, Jules....................... 94
Verrill, Addison Emery .. 191, 193, 194
Vikings.............................. 79
Visočica....146, 147, 148, 149, 151
Visoko......................146, 147
Vostok (rocket program) 9, 10
Walkabout (magazine).... 100, 101, 103

walking fish 21
Wallington, Wally .. 178, 179, 182
Wal-Mart................. 234, 236
Waters, Mel 35, 37, 39
waterspout.................... 19, 20
Webb, Dewitt 191, 193
Weiss, Philip 92
Welch, George......... 200, 201
Welles, Orson 122
Wikipedia.. 51, 180, 206, 208, 209, 210, 214
Williams, Montel... 40, 58, 64
Wilson, Sam (Uncle Sam) 127
Winfrey, Oprah. 63, 217, 219, 242
Woodbridge (Royal Air Force base)............. 105, 107, 111
World Health Organization (WHO).................... 16, 17
World War II.... 6, 24, 25, 27, 28, 162, 195, 196, 198, 253
Yeager, Chuck. 195, 196, 197, 199, 200, 201, 202
young earth creationism ... 165
YouTube ... 10, 15, 51, 56, 88, 110, 159, 178
zero-point energy............... 48
Zeta Reticuli 65, 69

CPSIA information can be obtained at www.ICGtesting.com
Printed in the USA
BVOW042209030912

299505BV00008B/14/P